Cognitive Sensors and IoT

Architecture, Deployment, and Data Delivery

Cognitive Sensors and IoT

Architecture, Deployment, and Data Delivery

Fadi Al-Turjman

CRC Press
Taylor & Francis Group
Boca Raton London New York

CRC Press is an imprint of the
Taylor & Francis Group, an **informa** business

CRC Press
Taylor & Francis Group
6000 Broken Sound Parkway NW, Suite 300
Boca Raton, FL 33487-2742

© 2017 by Taylor & Francis Group, LLC
CRC Press is an imprint of Taylor & Francis Group, an Informa business

No claim to original U.S. Government works

Printed on acid-free paper

International Standard Book Number-13: 978-1-138-10229-3 (Hardback)

To my dearest parents, my brother, and my sisters.
To my beautiful wife and my little prince.

Contents

Preface

Smart environments, enabled by Wireless Sensor Networks (WSNs), have emerged as one of the most promising applications in the Internet of Things (IoT) era. These smart environments require that the underlying sensor network infrastructure be enriched with smart devices, so that the network can understand and respond to requests from multiple users with diverse information requirements. Now, the use of artificial intelligence has enabled some amount of user-requirement awareness in sensor networks. However, there is no architectural framework about how cognition is to be incorporated in the network, or where the smart decision making shall be implemented. In addition, WSN implementations are mostly address-centric, where users must specify the location from where data must be gathered. But this is counterintuitive to how users would like to access information in a smart environment. Moreover, managing the large IP address space becomes problematic as the network size expands to vast counts of sensor nodes.

In this book, I propose a novel framework called *cognitive information-centric sensor network* (ICSN) for the IoT. This framework is built on top of cognitive nodes, capable of knowledge representation, learning, and reasoning, along with an information-centric approach for data delivery, which are central to the idea of the ICSN. Moreover, we discuss the most appropriate deployment strategy for these cognitive nodes under realistic assumptions that cares about the quality of information (QoI). In addition, a quality of information (QoI) aware data delivery strategy, with the analytic hierarchy process (AHP) as the reasoning technique, is used to identify data delivery paths that dynamically adapt to changing network conditions and user requirements. Latency, reliability, and throughput are the attributes used to identify the QoI along the delivery path. Furthermore, heuristic learning techniques are explored to improve the success rate of data delivery and hit ratio which applying in-network caching.

MATLAB® is a registered trademark of The MathWorks, Inc. For product information, please contact:

The MathWorks, Inc.
3 Apple Hill Drive
Natick, MA 01760-2098 USA
Tel: 508 647 7000
Fax: 508-647-7001
E-mail: info@mathworks.com
Web: www.mathworks.com

About the Author

Dr. Fadi Al-Turjman earned his PhD degree in computing science from Queen's University, Kingston, Ontario, Canada, in 2011. He is a leading authority in the areas of smart/cognitive, wireless and mobile network architectures, protocols, deployments, and performance evaluation. His record includes more than 120 publications in journals, conferences, patents, books, and book chapters. He has given numerous keynotes and plenary talks at flagship venues, including the IEEE ICC, LCN, GLOBECOM, and IWCMC conferences. Dr. Al-Turjman is also a visiting associate professor in the department of computer engineering at METU Northern Cyprus Campus. He has received several recognitions and best paper awards at top international conferences, and has led a number of international symposia and workshops in flagship ComSoc conferences. Since 2007, Dr. Al-Turjman has worked on wireless sensor network (WSNs) projects related to remote monitoring, as well as Smart Cities–related deployments and data-delivery protocols using integrated RFID-Sensor Networks (RSNs).

1

INTRODUCTION TO COGNITION IN IOT

The Internet of things (IoT) has evolved from supporting application-specific deployments, such as species-at-risk and health care monitoring, to platforms that simultaneously support multiple applications operating in a large-scale fashion, such as the smart planet project launched by IBM. In the IoT paradigm, every object in the physical world is uniquely identifiable and traceable, and can communicate with other machines and objects over the Internet. The use of sensors and actuators embedded into these objects helps to capture significant information from the physical environment and bring it into the digital realm over the Internet infrastructure. This interconnection between the physical and virtual realms paves the way for several information-centric applications and services that benefit from such interconnection. These applications are expected to support multiple-users accessing information from a city- or a planet-wide deployment of sensors that capture data from multiple applications. Users in this paradigm may target information for their personal use, or to build and support a large-scale database, such as the cloud.

Such a smart trend in communication that serves multiple users and multiple applications on the same platform necessitates a set of interconnected smart/cognitive nodes to carry on massive amounts of exchanged data over a resource-limited infrastructure. Cognitive nodes can communicate with the applications, the sensor network's nodes, and with each other to gather requested information and transmit it to the central data sink with the minimum cost.

Wireless sensor networks (WSNs) can be a strong candidate for employing cognitive nodes in smart environments because of their inherent ability to observe information from the environment and communicate it wirelessly to end users in a cost-effective manner. However, what is not well-defined yet is how these cognitive nodes

1

will be designed and integrated with the existing information-centric networks or infrastructures.

Accordingly, the objective of this book is to develop a comprehensive framework spanning the network architecture design, deployments issues, and the data delivery aspects. It aims to introduce the cognition in a well-known information-centric network paradigm, namely the WSNs, to be applicable in smart environments in the IoT era. In this framework, the end user must be able to search/ask for information based on the information attributes, rather than the IP address of the hosting machine. The network itself must be able to dynamically adapt for the varying network conditions, and make intelligent decisions about the data delivery routes to be used.

1.1 Contributions

In this book, we propose and implement the cognitive information-centric sensor network (ICSN) architecture for data delivery in large scale IoT applications. We assume a typical wireless sensor network (WSN), within which cognitive nodes are introduced. These cognitive nodes combine the use of cognition and an information-centric approach, to make data delivery decisions that dynamically adapt to user requirements. Towards this end, our main contributions in this book can be summarized as follows:

1. We start by defining the cognitive ICSN paradigm and related works in the literature. We identify critical factors and explore various learning techniques that can be applied to sensor networks in IoT applications. We provide our recommendation on the learning strategy based on how well it complements the needs of an ICSN, while keeping in mind the cost, computation, and operational overhead limitations.

2. We combine the use of cognition and information-centric networking to propose a novel cognitive ICSN architecture which can potentially support multiple applications in the IoT era. This architecture supports and resolves key IoT design aspects related to mobility and numerous data-traffic types.

3. According to the proposed ICSN architecture, we investigate the deployment planning problem while considering a new component in the network, called the *cognitive node*. We propose a virtual grid-based deployment planning strategy which guarantees the cognitive node connectivity with at least one sensor/relay node over the network lifetime span. Simply, the deployment plan is based on the probability of a successful reception of the data packet at a given distance from the transmitter.

4. We analyze and quantify the network connectivity of the aforementioned grid-based deployment planning under realistic operational conditions. Solid recommendations and endorsements based on analytical studies are discussed and debated.

5. We investigate the most appropriate data delivery techniques in the IoT era. We make use of the quality of information (QoI) metric to measure the level of satisfaction experienced by the end user. Attributes for the QoI, such as latency, reliability, and throughput are employed to assess the delivered information by the aforementioned cognitive ICSN architecture. Accordingly, delivery paths are cognitively identified using an analytic hierarchy process (AHP) based on the data-traffic type.

6. We propose two heuristic data delivery approaches that can take benefit of the defined cognitive elements in this work to cope with the next generation IoT trends. The proposed data delivery approaches either help to choose paths that deliver data with good QoI to the sink, or identify data delivery paths that are more resource-aware toward more energy- and cost-efficient solutions. The impacts of employing cognition on the success rate of the delivered data to the sink and the network lifetime are studied.

7. Moreover, a novel in-network caching approach is proposed to improve the hit ratios in an IoT-based information-centric network, where requested data can be intelligently cached over the appropriate nodes in the network based on the information value. This approach shows significant enhancements in terms of the network lifetime and data delivery latency.

1.2 Book Outline

The rest of this book is organized as follows. In Chapter 2, we delve into an overview for the field of cognition in ICSNs. In Chapter 3, we provide the details of our proposed ICSN architecture while discussing a novel grid-based deployment plan for the cognitive nodes. Chapter 4 provides the detailed analysis of the proposed grid-based deployment plan while considering practicality in surrounding environments. Chapter 5 investigates the details of the cognitive elements while employed in a generic data delivery ICSN framework. In Chapter 6, we provide the details of a cognitive energy-aware data delivery approach in the IoT era, which work with the learning technique to improve the average data delivery rate. We also compare the performance of the cognitive data delivery strategy against other commonly used techniques in the literature. In Chapter 7, we provide the details of another cognitive data delivery approach which focus on the cost in the IoT era. In this chapter we provide a pricing model for IoT-specific paradigms to improve the end user satisfaction. Chapter 8 debates the effects of caching in the aforementioned cognitive ICSN architecture. Finally, we conclude this book in Chapter 9 with potential perspectives on the proposed work and discuss the directions of the future work.

2

INFORMATION-CENTRIC SENSOR NETWORKS FOR COGNITIVE IoT

An Overview*

2.1 Introduction

Wireless Sensor Networks (WSNs) have evolved from simple sensing and tracking applications to being an integral and essential part of the Internet of things paradigm. This means that sensor networks have to deal with large amounts of data, support requests from multiple users, and support information extraction from the network rather than serving as point-to-point communication networks that transmit data from a source to sink. To enable WSNs to easily integrate with and adapt to the IoT environment, we propose the use of learning as an element of cognition in the network. Cognition refers to the ability to be aware of the environment, be able to learn from the past actions and use it to make future decisions that benefit the network [1]. Learning is one of the elements of cognition that can achieve different goals in different systems. In robotic chess, learning can be used to *plan* moves based on opponents' actions; in aircraft autopilot systems, learning can be used to *control* the plane's navigation, and in cognitive networks, learning can be used to improve *decision making* that improves network management and its overall performance. Whatever be the system's goal, the performance of the learning technique depends on three main tasks: (1) observations made from current activities in the

---

* This article was originally published in *Annals of Telecommunications*. F. Al-Turjman, Information-centric sensor networks for cognitive IoT: An overview, vol. 72, no. 1, pp. 3–18, 2017. Reprinted with permission.

environment, (2) feedback from past actions, and (3) how this information is used to achieve the system's goals. An information-centric sensor network (ICSN) has specialized nodes called *cognitive nodes* that are capable of performing all these tasks by implementing major cognition elements. These major elements are *learning*, *reasoning*, and *knowledge-representation*. Figure 2.1 represents these three major elements for cognitive nodes in ICSN, and associates them with their respective functions. These elements of cognition, when incorporated in the network nodes of a WSN, help in providing better understanding and catering for the end user requirements.

Our expectation from the cognitive elements would be to cater to the following objectives. In the short-term, to observe current network behavior and respond adaptively to changing network dynamics. And in the long term, to learn from the previous behavior and plan better for the future so as to make predictions and decisions that positively impact the network survivability and application QoI during its lifetime. Using these elements, a conceptual architecture of the cognitive node can be illustrated as detailed in Figure 2.2.

This cognitive sensor network framework is mainly targeted toward applications, such as Smart Cities and Smart Outdoor Monitoring in the IoT era. In these applications, the goal for the learning algorithm is to dynamically adapt routing decisions to improve the quality

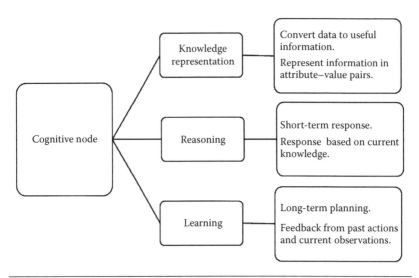

Figure 2.1   Cognitive node and its elements.

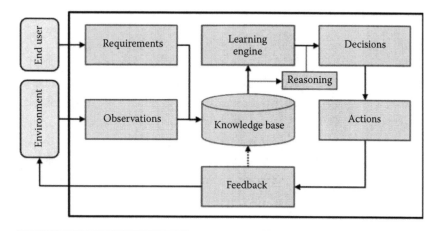

Figure 2.2 Cognitive node conceptual architecture.

of information (QoI) delivered to the user [2], and improve the network lifetime. This can be achieved by using a learning algorithm that can correlate the impact of past decisions on the current network behavior, based on end users' requirements. But these are challenging applications for the learning algorithm due to the large scale of the network, changing network topology due to node deaths and varying channel conditions, and heterogeneous traffic flows generated as a result of changing user requests in the IoT application domain. For the learning algorithm to be successful in such a dynamic environment, it must occur incrementally and span over multiple episodes [3]. Only then will it be able to adapt to the irreversible changes in the environment and contribute toward delivering QoI-aware data to the end user over the network's lifetime. In this chapter, we explore the different classes of learning techniques and identify what works best for large-scale information-centric sensor networks in IoT applications.

The remainder of the paper has been organized as follows: In Section 2.2 we look at some WSN design issues in IoT applications, and summarize the design changes required for integration of WSNs with IoT. In Section 2.3 we explore artificial intelligence and learning used by WSNs to better analyze their suitability for ICSN applications. Subsequently, we delve into the details of ICSNs as a hybrid solution platform for integration of WSNs in IoT using a learning paradigm in Section 2.4. We also identify a suitable learning strategy in this section, before discussing some open issues and concluding the paper in Section 2.5.

2.2 WSN Design Issues in IoT Applications

In this section we take a brief look at the various design issues that need to be addressed to seamlessly integrate WSNs in IoT applications. We categorize the design issues into two parts: (1) Expectations of the users from the network, which includes feature requirements of the sensor network's interface with the access network (future Internet), and (2) adaptations and changes required within the sensor network to cater to user requests while managing network resources. Each of these design issues will be explored in the subsequent sections, and we will see why these issues need to be handled differently from existing WSN applications.

2.2.1 User Expectations from the Network

Traditional WSNs were designed for specific applications such as target tracking, temperature monitoring in a building, and movement of goods in a supply chain, to name a few. Users accessed the network only when they needed a particular type of sensed value, such as temperature, pressure, or humidity for instance. However, in the IoT era, sensor nodes have become heterogeneous and are capable of supporting multiple types of sensors. This way the sensor network can be expected to support multiple applications and provide users with a variety of data as supported by the type of sensor nodes used. Thus, WSN applications should evolve from supporting application-specific deployments to providing an application platform that users can access to gather a variety of data [4].

2.2.1.1 Multi-User Application Platform Support   The basic idea behind developing an application-platform is to provide a flexible, generic infrastructure that can lead to easy adoption of WSNs into a variety of IoT applications. For example, a sensor network deployed in a city should be capable of providing data for the following applications: (1) air pollution monitoring, (2) daily weather monitoring (temperature, humidity, UV index), (3) park and garden irrigation management. Such an application platform and its associated services would also support the conceptual ubiquitous sensor network (USN) used in large-scale sensor network deployments [5,6]. This would invite more numbers of users to simultaneously access the network, which makes the WSN design even more complex. The underlying sensor network will have to support heterogeneous traffic flows generated as a result of simultaneous access from multiple users, which is a very challenging task for the resource constrained sensor network.

2.2.1.2 User Requirement—Aware Request Classification   In a large-scale deployment of sensor networks that allows multiple users to access it, different users may have different requirements on the desired quality of experience (QoE) [7], and the network may have its own limitations on the quality of service (QoS) it is able to support [8–10]. While the user requirements may evolve over time, the sensor network gradually decays in terms of its energy capacity, and it also involves dynamic changes in the link conditions and node availability. In addition, the user's expectations from the attributes associated with delivered data also vary based on the application and type of request. Hence, there is a need to monitor data attributes from the user requirements perspective directly, a skip level from the application interface, as shown in Figure 2.3. Latency, reliability, accuracy, relevance, and robustness are some of the attributes that can collectively provide an estimate of the quality perceived by the user based on the information received at the user-end from the network. This is referred to as the quality of information (QoI) metric, and it may provide information about the success of the network in satisfying the evolving user requirements, while simultaneously saving valuable network resources such as bandwidth and energy [11,12]. Thus, it becomes necessary for the underlying sensor network to be aware of user requirements and be able to classify user requests to deliver data in compliance with the desired QoI. While there has been research in the area of quality of information (QoI) assurance based on changes in the phenomena observed in

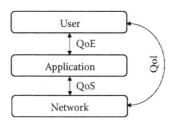

Figure 2.3   Quality of information–aware data delivery in WSNs for IoT applications.

the network [13], there has not been much work on the user-interface side for request classification based on changing user requirements.

2.2.1.3 Internet Access Interface between User and WSN   ZigBee-based address-centric sensor networks facilitate the integration of WSNs with the Internet, as both the networks are address-centric and point-to-point. However, recent advances in the future Internet research suggest that an information-centric approach to data delivery is favored over the address-centric approach. This is because researchers believe that it will become increasingly difficult to handle the growing IP address space that serves billions of users in the near future. Information-centric network (ICN) is one of the promising architectures for the future Internet paradigm [14]. It proposes an information-centric approach to data access where users look for named data objects instead of looking up IP addresses to find data they are interested in. The architecture is set up to support data storage at strategic locations in the network, typically the edge of the network, so that requests do not have to be propagated deep into the network to access the required information. If the user had to request for data through such an information-centric interface, then the network interface would also have to be modified to ease the flow of information. Hence, for IoT applications to be able to adapt to changes of the future Internet paradigm, changes would be required in the way requests are made at the user access interface and also at the sensor network interface.

2.2.2 Adaptations at the Network Level

In this section, we move from the issues at the user interface to the design issues at the network level to identify the adaptations required to enable WSNs for IoT applications.

We look at the adaptations required in WSN for IoT applications from the perspective of network resource conservation, and managing the query dissemination and data delivery from the network.

2.2.2.1 Energy Considerations and Resource Management   Large scale platform-based sensor network deployments that are accessed by multiple users tend to benefit more from a data-centric approach than an address-centric one. For instance, a user may request information such as temperature readings from all regions in the network where value is greater than 25°C. In such a scenario, there is no specific address from which the user is requesting data. Instead, the query is information-centric, and requires information from the entire network. This could be very energy-intensive if appropriate query dissemination and data delivery techniques are not identified. In their work on energy conservation schemes in WSNs, Anastasi et al. have extensively explored data-driven techniques, broadly classified into data reduction and energy-efficient data acquisition approaches [15]. Several researchers have also established the energy conserving nature of data-centric sensor networks [16–18]. However, previous research has only considered networks of a few hundreds of sensor nodes. In IoT based applications, there may be thousands of sensor nodes to gather information from, which adds to the complexity of the energy conservation problem. Although a data-centric approach can provide valuable energy savings in the network, further research is required to devise techniques to manage information flow in such a large scale, energy constrained network.

2.2.2.2 Query Dissemination and Data Delivery WSNs can be broadly classified into address-centric and data-centric networks. This classification is based on how query dissemination, data gathering, and routing happen in the network. Address-centric sensor networks are built on top of the more recent ZigBee protocol [19] that provides a service-oriented framework for implementing WSN applications. Data is routed using the tree-based hierarchical topology consisting of router and coordinator nodes, while sensor nodes are the sources of information. Routers off-load the sensor nodes by carrying forward their data to the sink, thus bringing considerable energy savings to the battery-operated sensor nodes. However, WSNs are essentially information extraction networks. They were originally developed as data-centric sensor networks (DCSNs) that did not make use of node addresses. Instead, their focus was on the attributes of the requested information, which was gathered from wherever it was available in the network, and delivered to the sink. Handling query dissemination and delivering the gathered data is a very challenging task in large-scale sensor networks. This is due to (1) the ad hoc nature of the wireless channel, (2) dynamic topology changes in the network due to node deaths and their changing associations, and (3) the nature of node distribution in the network. Hence, choosing the right approach for handling data flows in a very critical design decision; this must be made keeping in mind the interface access network (the future Internet paradigm).

In addition, IoT applications have multiple users requesting different types of data with different service requirements. For example, while requests for periodic data may relax the service requirements on latency, on-demand data needs to be provided quickly, and should be relevant. On the other hand, emergency reporting must be done accurately, reliably, and quickly, with minimal delay. Thus, the way the network is set up itself will have to be modified to minimize energy consumption during each of the phases of query dissemination, data gathering, and data delivery. Sensor node scheduling also becomes an important issue to be addressed, so that data is available when the user requests it. Moreover, planned scheduling of sensor node wake-up and sleep cycles will add to the energy savings and prolong the network lifetime. In addition, planning the deployment of router nodes to increase the multi-hop communication range is another aspect that needs to be considered during network design and deployment. Since the complexity of the tasks to be handled by WSNs in the IoT paradigm is quite high and multidimensional, it seems appropriate that the design changes consider the addition of advanced nodes [20] that can maintain connectivity in the network and carry data over long distances in these applications.

2.2.3 Summary of WSN Design Change Requirements

Thus far in this section, we have seen the various design change requirements in existing WSNs to make them adapt to IoT applications. We summarize these requirements as follows:

- Multi-user application platform support.
- Classification of user requests to deliver QoI aware data.
- Modification of the communication to make it compatible with the future Internet paradigm—ICN.
- Incorporation of specialized nodes that can observe and learn from the interactions and feedback in the network, and manage sensor node scheduling to prolong the network lifetime.

- Plan the deployment of router and specialized nodes to maintain network connectivity and enable multi-hop data transmission over the large-scale network.
- Consider data-centric query dissemination and data-delivery for energy savings, and dynamic traffic flow management due to changing network conditions and user requests.

Figure 2.4 summarizes all the design change requirements in the form of a conceptual design for the future IoT paradigm that supports multiple users, and integrates ICN based Internet access, and the large-scale data-centric sensor network. We call this an *information-centric sensor network* (ICSN) [21–24]. Comparing DCSN protocols with the information-centric networking (ICN) approaches [25,26] for future Internet use, we can see that DCSNs already implement two major features of the ICNs. Firstly, the naming schemes, or named data objects for referencing requested data instead of using node addresses, and secondly, storage of collected data in nodes for ease of access. We take these two features as a strong indication of the need to shift to data-centric sensing schemes for WSNs, but with ZigBee and the information-centric approach to adapt to the advanced applications in the IoT era [27]. Although the data-centric approach will help to

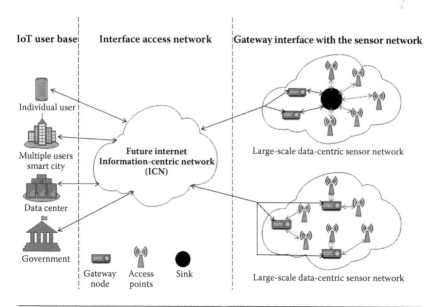

Figure 2.4 Conceptual design of an information-centric sensor network for IoT applications.

better manage the network at each of the interfaces: User, access network, and sensor network, the author does not think that a simple integration of different technologies will be adequate to manage the complexity of the tasks at the sensor network level. To improve adaptations beyond the limitations of traditional cross layer design, we look to the tools of artificial intelligence to support the ICSN paradigm. In particular, we look at how ICSNs can benefit by incorporating the learning aspect of artificially intelligent systems in the next section.

2.3 Artificial Intelligence and Learning in WSN

Various artificial intelligence (AI) techniques have been applied to WSNs to improve their performance and achieve specific goals. We look at AI techniques as a means of introducing learning in the WSN. Learning is an important element in the observe, analyze, decide, and act (OADA) cognition loop [28,29], used to implement the idea of cognitive wireless networks [30,31]. In this section, we broadly classify AI techniques as computational intelligence (CI) techniques, reinforcement learning (RL) techniques, cognitive sensor networks and multiagent systems (MAS), and context-aware computing, as shown in Figure 2.5. Although these techniques are closely related, we segregate them to show the different goals that learning can achieve for the network. At the end of this section, we look at how these techniques could be improved and leveraged to support learning in the ICSN platform.

2.3.1 Computational Intelligence

CI techniques are a set of nature-inspired computation methodologies that help in solving complex problems that are usually difficult to fully formulate using simple mathematical models. Examples of CI techniques include genetic algorithms, neural networks, fuzzy logic, simulated annealing, artificial immune systems, swarm intelligence, and evolutionary computation. In a learning environment, CI techniques are useful when the learning agent cannot accurately sense the state of its environment.

In WSNs, CI techniques have been applied to problems such as node deployment planning, task scheduling, data aggregation,

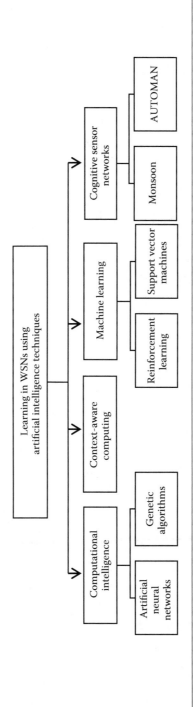

Figure 2.5 Learning techniques used in WSNs.

energy-aware routing, and QoS management. Kulkarni et al. have provided an extensive survey of CI techniques applied to WSNs [32]. They elaborate on various CI techniques, and associate each with typical problem domains they can solve in WSNs. From their observations, swarm intelligence applied to solving the routing and clustering problem has drawn the most research attention in recent times. However, a major drawback of this methodology is that it can be computationally intense and may require some form of model-based offline learning to deliver to the requirements of the application scenario. Techniques such as ant colony optimization can cause an undesirable increase in communication overhead in WSNs too [33]. Apart from these drawbacks, none of the CI algorithms have been applied to solving problems of data representation, aggregation, and delivery in a distributed, decentralized setup, under dynamic communication constraints, as is the case in IoT applications. Next, we evaluate reinforcement learning strategies used in WSNs. We separate them from other CI techniques, because they are the most widely applied learning techniques in WSNs and do not categorically fall under nature-inspired learning.

2.3.2 Machine Learning

Machine learning can be classified into supervised, unsupervised, and reinforcement learning. Supervised learning would be more computer-intensive and require a training sequence. Additionally, accuracy of the learning algorithm would then be defined by this training sequence. In the unsupervised learning approach, the learning is from the environment being observed and no training sequence is required. Reinforcement learning (RL) is a reward-based technique that emphasizes learning while interacting with the environment, without relying on explicit supervision or a complete model of the environment. It is a method of automating goal-directed learning and decision making. Since it is advantageous to be aware of, and learn the changes in the environment in WSN based applications, RL would be an appropriate choice as the learning strategy. In WSNs, RL has been successfully applied in networking tasks such as adaptive routing; identifying low cost and energy-balanced data delivery paths [34,35]; and in information processing tasks involving data aggregation and inference [36]. In the subsequent sections we explore the different types of reinforcement learning (RL) methods as reward based strategy [37]. We also briefly look at the use of support vector machine techniques as an RL strategy that the ICSN could benefit from.

2.3.2.1 Model-Based Reinforcement Learning   A learning agent (LA) in model-based RL collects experiences and builds a model from that. The actions can be chosen randomly or heuristically and observe the impact on a model $T(s, a, s')$ and the reward $R(s, a, s')$ (where s is the current state, a is the actions and s' is the next state). This means, by taking action a, how often would the LA end up in the state s' if it started from state s and estimating a probability by counting the number of times $T(s, a, s')$ triple occurs over the sample space. With this information, the LA builds an estimate of the model $\hat{T}(s, a, s')$ and the reward $\hat{R}(s, a, s')$ by estimating probabilities based on the number of trials (or episodes). Once an estimate of the model and rewards are ready, the LA can plan its actions. A good plan can be found from policy iterations or value iterations. But for model-based RL, the hardest part is knowing the right policy to start out with, so we can build a good model.

2.3.2.2 Model-Free Reinforcement Learning   In model-free reinforcement learning, the agent is free to learn from the environment by exploring it completely on its own. The agent learns from the positive reinforcement it gets for moving toward a goal and negative reinforcement for moving away from the goal. Q-learning is a form of model-free RL in which the learning agent converges to an optimal policy even if it were acting suboptimally. This is called off-policy learning. This is the most extensively used form of RL, as it is easy to implement and a relatively low-cost solution. However, Q-learning has its limitations too. The agent has to explore enough and eventually make the learning rate small but not decrease it too quickly, so that it has a large enough state space that covers all possible actions and policies.

In the context of the ICSNs, we are not interested in finding the optimal policy. Rather, we are interested in any suboptimal solution that does not take much time to converge and can act faster. It doesn't even matter how the action is selected, and a heuristic choice could work well too. This way, an even more simplified version of RL can be applied to ICSNs. We will discuss this in more detail in Section 2.4.

2.3.2.3 Support Vector Machines   Support vector machines (SVMs) are a class of ML algorithms that were originally formulated for binary classification. They can be applied to complex, highly nonlinear problems in a consistent and structured manner, and have been successfully applied to intrusion detection and security problems in WSNs [38]. This technqiue has not been applied to any other design problem in WSNs, but has proved very effective in small-scale sensor networks of about 50 sensor nodes randlomly deployed in a 100 m × 100 m area. This class of ML algorithms has great potential to be applied to security in IoT based applications, especially if the interface becomes information-centric, with gateway nodes lying within the sensor network itself, as shown in Figure 2.3. As the sensor network becomes more vulnerable to attack in the ICSN setting, SVM techniques can be further explored to secure large-scale networks in IoT applications.

2.3.3 Cognitive Framework and Multiagent Systems

Cognitive networks are built around the idea of having sensor networks evolve around user requirements. It is about taking a step toward developing intelligent networks that do not limit themselves to point-to-point communication within the network. Instead, they enable the network to perceive user requirements and deliver data using distributed intelligence in the network. To implement distributed intelligence, multiagent systems (MAS) are typically used. The agents in these MAS are called cognitive agents. They may interact to achieve information fusion and retrieval, and may also be able to predict data for future use. Typical applications include medical monitoring, intelligent transport, and smart environment monitoring [39]. Such a distributed approach to introducing intelligent behavior in the network will be very beneficial in WSNs deployed for large-scale IoT applications. Some examples of successful software implementations of cognitive agents in sensor networks include AUTOMAN [40] and MONSOON [41].

2.3.4 Context Aware Computing

In large-scale sensor networks, a huge amount of data is generated. In order to derive useful information from raw data, context of the data plays an important part. Context awareness is even more important in the IoT era, as it enables the network to deliver relevant, user-requested data. While doing so, network resources are also conserved by extracting only meaningful information that is relevant to the requests from the network. There are various aspects to context aware computing. They are acquiring the context, context modeling, context reasoning, and context distribution [42]. Since we are more interested in the learning aspect as applicable to sensor networks, we delve a little deeper into the context reasoning aspect. Both CI- and RL-based techniques can be used to implement context learning, in addition to ontology based and probabilistic logic models. The information obtained from these learning models can be exploited to infer information from stored data. Context awareness is very important and valuable in IoT applications, as it can add value to the large amount of data available from their applications.

All the learning techniques discussed in this section have been summarized in Table 2.1. We identify the limitations of existing techniques and provide the details of the solution platform based on ICSNs for IoT applications in the next section.

2.4 A Hybrid Solution Platform: Learning in ICSN

In this section, we will identify what the learning algorithm should actually learn in order to support the information-centric nature of ICSNs. Table 2.2 presents three broad classes of solutions to RL problems and classifies their features based on their relevance to learning in ICSNs.

2.4.1 What Should the ICSN Learn?

To identify what the ICSN should learn, let us identify what information is already known to the learning agent nodes. First, from our work in [22] and [43], we know that there is the upper bound on the maximum communication range for all the network nodes—sensor

Table 2.1   Learning Using AI Algorithms in WSNs

	TECHNIQUES USED	DESIGN PROBLEMS ADDRESSED	ROLE OF LEARNING	LIMITATIONS
Computational Intelligence Techniques	Neural networks Genetic algorithms Fuzzy logic Simulated annealing	Node deployment Power assignment Routing Network management (coverage, lifetime)	Offline learning phase Learning to estimate congestion based on packet arrival rate and buffer size Buffer threshold management by learning packet delay and throughput values	Difficult to define appropriate quality measures for Pareto set approximations Poor adaptation to changing network topology due to offline learning phase High cost of operation (energy and delay)
Machine Learning Techniques	Reinforcement Learning Q-Learning and its variants (FROMS) Support vector machines	Routing Node deployment Resource allocation Intrusion detection network security	Online learning of route costs Energy efficient data aggregation paths Outlier detection	Very application-specific design choices prevent building upon knowledge gained from learning in different applications Long time to converge to solution
Cognitive framework and Multiagent Systems (MAS)	AUTOMAN (intelligent software agents) MONSOON	QoS management Surveillance applications Distributed intelligence for data collection under dynamic network conditions	Event-driven learning to capture changes in specific parameters (voltage) on the power grid in AUTOMAN Learning data delivery paths based on pheromone trails, adapting to node and link dynamics	Application specific design Policy-based design, no generic infrastructure Cost of internode communications among intelligent agents is high

(Continued)

Table 2.1 (Continued)   Learning Using AI Algorithms in WSNs

	TECHNIQUES USED	DESIGN PROBLEMS ADDRESSED	ROLE OF LEARNING	LIMITATIONS
Context Aware Computing (context learning)	Supervised Unsupervised Rule-based Fuzzy logic Ontology-based Probabilistic logic	Activity or event recognition Unusual behavior detection Automated irrigation when soil is "dry"	For context-based reasoning and modeling decisions Using previously stored knowledge to infer contextual knowledge	Context modeling might be complex No validation or quality checking Each of the context learning techniques have unique limitations
ICSN	Knowledge representation Learning Reasoning	User requirement awareness and request classification QoI-aware data delivery Distributed intelligence	Predicting user requirements and request types Learning from feedback and observations in the network to improve QoI along data delivery paths Can be used for implementing context awareness by processing gathered data	Conceptual design that has not yet been validated in experiments Cognitive nodes might be more expensive compared to relay nodes Heuristically accelerated RL can help to reduce the learning time

Table 2.2 Classification of Solutions to RL Problems

	FEATURES	INCREMENTAL LEARNING SUPPORT	REQUIRES MODEL OF ENVIRONMENT	BOOTSTRAPPING (LEARNING A GUESS FROM A GUESS)	LIMITATIONS	EXAMPLES
Dynamic Programming	• Computes optimal policies using a perfect model of the environment • Estimates of values of states are updated based on those of successor states	No	Yes	Yes	• Computationally very expensive • Not suitable for large problems	Policy improvement, Policy iteration, Asynchronous-DP
Monte Carlo Methods	• Requires online or simulated sample of interaction with environment for state transitions • Averages the values of sample returns	Yes	No	No	• Maintaining sufficient exploration is an issue	On-policy methods, Off-policy methods
Temporal Difference	• Make long-term predictions about dynamic systems • Can be applied online with minimal computation	Yes	No	Yes	• No notable limitations; perform more efficiently with eligibility traces	Q-Learning, Actor-critic methods

nodes, relay nodes, and the nodes that implement learning as a part of cognition, the cognitive nodes. We will be referring to the learning agents as cognitive nodes (CN) from this point forward. So, the CNs need to be aware of the network topology changes only within their own communication range. The target area's coverage is taken care of at the time of deployment of nodes in the network.

Thus, *learning about the topology changes in its local neighborhood* will help the CNs adapt their transmit power and choose a data delivery path that best manages the *nodes energy consumption*. Prolonging the CNs lifetime will in turn contribute toward *increasing the network's longevity*.

Second, cognitive nodes store information in their knowledge base regarding the QoI performance of paths used in previous data-delivery rounds. Routing tables are built and updated based on the information in the knowledge base. Unlike traditional routing tables that store static, end-to-end routing paths from source to destination nodes (usually the sink node), routing tables in ICSNs are not designed to be static. In fact, they are not even end-to-end paths, but are paths that show the network's current adaptation to the changes in topology and user requests. They store information about the most recent path used to deliver data from the CNs to the sink or other CNs or relay nodes. This means that the contents of the Routing table at each cognitive node have to be updated on a regular basis to ensure they store the latest and best QoI paths. Thus the important learning goal for the cognitive node from an application QoI perspective *is to learn the data-delivery paths toward the sink that provides the best QoI values for each of the different types of user-requests*. There is no one best path that is always used to route data. Instead, the routing choice depends on the current network topology, nature of the user-request (periodic, intermittent-user specific, or emergency data), and volume of traffic generated in response to the request. If the routing table is viewed as cache storage, then an effective cache replacement strategy is required to replace old and redundant routing information with more recent and relevant information.

These learning goals can be achieved in the following ways as represented in Figure 2.6: (1) learning from feedback on current actions, (2) learning by exploring the changes in the network topology, and (3) learning (drawing inference) from past actions by using the

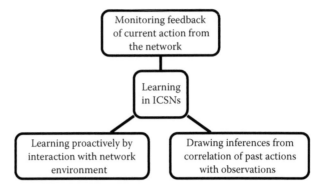

Figure 2.6   Learning in sensor networks.

information stored in a Knowledge Base. Once the learning converges and knowledge base updates are complete, the routing tables can be updated using information from the knowledge base. Thus we identify two main goals of learning: First, to improve network longevity and maximize the duration for which the ICSN is able to provide relevant data to the end user; and second, to increase the network's ability to provide QoI-aware data to the end user during its lifetime. For ICSNs, we define the network lifetime as the time t_{EoL} after which none of the one-hop cognitive nodes to the sink are able to deliver the user's requests for information to the network, nor are they able to deliver the gathered data from the network back to the user, is called end-of-life of the ICSN.

2.4.2 Choosing the Learning Strategy

The goal of learning in ICSNs is to improve the QoI delivered to the end user and maximize the duration over which this can be maintained. In other words, learning should contribute toward maximizing the network lifetime while ensuring QoI-aware data delivery during this period. Based on these requirements, we identify the following factors that influence the choice of learning strategy in ICSNs:

- Learning should occur incrementally over time as network conditions change.
- It should support episodic learning [43] to acquire information from new events and update the knowledge base. This will let the system make decisions and act based on the observed changes.

- Learning should be distributed and occur locally when implemented in large-scale ICSNs. However, the goal of learning should be common across all the cognitive nodes in the deployment.
- The learning algorithm must be lightweight and of low-complexity to support episodic learning. It addition, it must not cause too much data or control overhead in the network or negatively interfere with the function of the network nodes.
- The learning algorithm should preferably be reward-based, as supervised learning would become very compute-intensive for the ICSNs.
- Finally, the learning algorithm must be able to converge quickly enough to support the network's learning goals in a timely manner. It should not cost the network too much in terms of its resources (energy and time) either.

Based on these recommendations, we explore heuristically accelerated reinforcement learning techniques for use with ICSNs in the next section.

2.4.3 Heuristically Accelerated Reinforcement Learning

Since both model-based and model-free RL have their limitations, we will look toward modified forms of RL solutions. We look at the possibility of using heuristics (rules of thumb) to choose a suboptimal action instead of trying to converge at an optimal policy. This would help the RL algorithm to converge faster. The advantage of combining heuristics with RL is that RL is eventually going to converge to an optimal possible policy, but it takes time. Heuristic only attempts to make the decision choices quickly. Therefore, it will not deter the RL process, but will only enhance it to arrive at some sub-optimal solution faster, but it will not be incorrect as observed by Atkeson and Santamaria [44]. In this section, we look at the use of heuristic functions to accelerate RL algorithms. These heuristic evaluation functions, called *valuation functions*, are computed by stochastic sampling and dynamic programming updates [45]. The model-free method is suitable for problems that do not involve large state spaces. In contrast, the domain-independent, model-based heuristic methods can be used for solving problems with a large state space and hundreds of actions. The fast and frugal heuristics proposed by Gigerenzer [46,47] are not only generic, but are also low-polynomial time and applicable to all problems that fit a given model. Another aspect to consider is that learning happens while the system is running. This makes it important to reduce the exploration space of the learning agent. In the CICSN framework, we can use the information stored in the knowledge base as a case base to choose an action that provides a close solution to the RL agent's problem.

But the choice of the case that matches the given decision problem must be chosen in a way that helps the RL algorithm to converge quickly. Thus we look at the possibility of making heuristic decisions to choose a case from the knowledge base that matches closest with a given decision problem and adopt the existing solution to find a new one that is relevant to the decision problem at hand [48]. This way we can limit the exploration space by making use of the knowledge base, and use heuristics to choose a suboptimal action that will help the algorithm converge faster. Heuristically Accelerated RL (HARL) and case-based HARL have been recently explored by Bianchi et al. in the context of robotic soccer [49,50]. We extend this idea of accelerating RL by using heuristics and an available case base to the ICSN framework. In our application, the case base is replaced by a Knowledge Base that built upon the foundation of representing knowledge in attribute–value pairs.

The Representativeness Heuristic: According to this heuristic, the more similar something is to a prototype, the more likely it is to belong to that prototype's category [51]. This heuristic is based on the fact that we tend to judge how likely something is to be true based on how representative it is of a particular category.

Thus we conclude this section with the suggestion that heuristically accelerated RL techniques that make use of the information stored in the knowledge base of the cognitive nodes will potentially serve as a low complexity solution to the learning problem in ICSNs, and might be viable in terms of the computational overhead, too.

2.4.4 Learning in the ICSN Framework

In the previous subsection, we identified a learning strategy for ICSNs based on the network requirements and from the analysis of different learning strategies. Now we will identify how the ICSN must be set up so that the learning strategy can be implemented in this framework. We assume a deterministic deployment of relay and cognitive nodes, and number them for ease of representation.

We start with a broad classification of the type of request that an ICSN should be able to serve. We divide the requests into one of three types: (1) Type I: Periodic, (2) Type II: On-Demand, and (3) Type III: Emergency request. Each of these requests will be associated with different QoI values on the delivered data, as desired by the user. We identify Latency (L), Reliability (R), and Throughput (T) as the three attributes, whose combined value will decide the QoI-associated with the delivered data. Energy efficiency is another important parameter that affects the network's performance and impacts the network lifetime, and we will consider it while making decisions in the network, especially when related with choosing a data delivery path. We will not use absolute values of these attributes in deciding the QoI value. Instead, we will associate priorities with each of these attributes for every request type, and let these priorities decide the importance of the absolute value of the attributes. Thus, each request type is classified according to the priorities associated with QoI attributes of L, R, T, and the importance of considering energy efficiency in making a decision choice in the network, as shown in Table 2.3.

The QoI attributes are monitored from feedback in the network. When a data packet is transmitted from a CN to its one-hop neighbors, the QoI attribute values are piggy-backed along with the acknowledgment it receives from these nodes. These values will be stored in a knowledge base (KB) and used in deciding the most appropriate next hop for subsequent requests arriving at that CN. This way, decisions about data delivery paths are dynamic and always based on both the user requirements and the network conditions at any point.

Table 2.3   Priority Associated with QoI Attributes for Different Request Types

REQUEST TYPE	QoI ATTRIBUTES			
	LATENCY	RELIABILITY	ENERGY	THROUGHPUT
Type I: Periodic	x	3	1	2
Type II: On-Demand	1	2	4	3
Type III: Emergency	1	1	x	2

The advantage of making these decisions at CNs is that it helps in decentralized decision making. Moreover, only local, one-hop neighborhood information needs to be monitored and stored. This means that the size of the KB to be maintained remains reasonable and can be easy to update and maintain. Table 2.3 shows the association of QoI attribute priorities with each request type. The numbers in the table indicate the priority associated with the attribute. Number 1 indicates top priority and number 3 indicates least priority. The "x" in the table indicates a "don't care" condition. This means that there are no strict requirements on the value of the QoI attribute marked with an "x," and its value does not impact the decision making.

Next, we look at the structure of the KB, where all the information gathered from observations in the network, and learned from feedback is stored. The KB thus stores all the relevant and useful information that the learning and cognitive decision making algorithms can use. It also serves as a case base which the learning heuristic can use to map a given problem with, and decide on the best course of action for the future. Table 2.4 represents the KB at a cognitive node, and has a sample of the information stored in it as attribute–value pairs. Attribute–value pairs are one of the techniques used for knowledge representation. Information is represented in a way that the user can derive useful information from it, by drawing inferences about how the values are connected [52]. The inferences drawn could be based on

Table 2.4 Sample of a Knowledge Base and Its Contents at the Cognitive Node

ATTRIBUTE		VALUE		
1-hop neighbors	ATTRIBUTE'	VALUE'		
	Node type	Sink	RN	CN
	Distance	400 m	250 m	350 m
	Remaining Battery	∞	200	300
QoI	Request Type	I	II	III
	Node	RN3	RN6	SINK
	QoI attributes	$L = \alpha_1$	$L = \alpha_2$	$L = \alpha_3$
		$R = \beta_1$	$R = \beta_2$	$R = \beta_3$
		$T = \gamma_1$	$T = \gamma_2$	$T = \gamma_3$
Sensor data	Temperature	25		
	Humidity	20		
	UV Index	5		
	CarbonMonoxide	250		

rules, or heuristics based on learning from observations and feedback in the network. In Table 2.4, a recursive representation of attribute–value pairs has been used. That is, each entry in the *value* field can be another attribute–value pair. What makes this representation effective beyond the attribute–value association is that information can be derived by reading the values along the column too, except for the field containing sensor data. For example, in the *Attribute* field, "Node type" is associated with three values: "Sink," "RN," or "CN." RN represents relay nodes that the CN is connected with and CN represents other cognitive nodes that the given CN is connected with in the ICSN. The *Distance* field corresponds with the values in the "Node type," and represents the separation between the CN housing the KB and the sink, RN, and CN, respectively. Tracking the remaining battery level at each of the one-hop nodes helps the CN take energy-aware decisions in choosing the data delivery path. The next major Attribute we have used is "QoI." It has information about the "Request Type" (as described in Table 2.3) that the node has served, next hop "Node" that can best serve each request, and the values recorded for each of the "QoI attributes" of latency (L), reliability (R), and throughput (T) during the previous communication. These values could be different between any pairs of nodes and are thus represented by α^*, β^*, and γ^*. It should be noted, however, that this table is only a representation of how information can be stored in the KB. In actual implementation, details of the semantics will have to be worked out to make the representations shorter and effective.

In the proposed ICSN framework, we even segregate the routing table from the KB to keep routing decisions simple. Routing tables at the CNs store information only about reaching the one-hop neighbors, not the end-to-end paths, as shown in Table 2.5. These entries are derived from the KB of Table 2.4. In Table 2.5, the "Possible next hops" field suggests the best next hop node for CN2 (Cognitive Node 2) to transmit data, based on the "Request type." It shows that CN2 is directly connected to the sink, connected to four RNs that are linked with the sink, and is also connected to two other CNs. CN–CN paths are not preferred and are represented by the hyphens in the "Request type" column. This is due to the high cost in terms of energy consumption, and the possibility of running into loops without reaching the sink. These tables can be updated every time the learning

Table 2.5   Routing Table at the CN

ROUTING TABLE FOR CN2	
POSSIBLE NEXT HOPS	REQUEST TYPE
Sink	III
RN6-> Sink	II
RN7-> Sink	II
RN2-> CN1	I
RN3-> CN3	I
CN1	–
CN3	–

algorithm identifies better paths for each request type, based on the changing network dynamics as reflected from the KB. In addition, a reasoning algorithm to help the learning agent in making cognitive decisions must be identified. These are still open research issues that need to be addressed in the future.

2.5 Use-Case and Performance Evaluation

In this section, we provide some performance evaluation of the different learning techniques in improving large scale ICSNs for Cloud- or IoT-based applications. As described in the context of cognitive psychology [48], the learning heuristics will be used as strategies that ignore a part of the information to make decisions faster, and sometimes more accurately compared to more complex methods [25]. We utilize an online version of the A* heuristic search algorithm, which learns from the information available in the knowledge base of the cognitive nodes. We call this Learning Data Delivery A* (LDDA*) algorithm. The heuristics will be used to make approximate decision choices, as opposed to optimal decision choices. We compare this with a cumulative-Heuristic accelerated learning (CHAL) technique that accumulates the heuristic values at each state (relay and cognitive nodes), and makes use of as much information as possible from observations made in the network before making the data delivery path choices. It also uses negative heuristic weights to punish poor next hop node choices, such as revisiting a node along a data delivery path. This way, LDDA* and CHAL will differ in the heuristic weights accumulated by the learning process. Since learning is typically used to improve the decisions made by the reasoning engine in cognitive networks, we implement LDDA* and CHAL in a network that uses an analytic hierarchy process (AHP)–based reasoning technique at the cognitive nodes to make data delivery decisions. (The details of AHP-based data delivery [AHPDD] have been described in our previous work [21].) Performance of the heuristically accelerated learning techniques LDDA* and CHAL are compared against the nonlearning AHPDD in terms of the QoI observed at the sink where data is delivered at the end of each transmission round. The algorithms will also be compared in terms of the rate of successful data delivery and the energy consumed during the data delivery process at the end of the network's lifetime. The knowledge of the deterministic deployment of the RNs and CNs, and the knowledge accumulated in the knowledge base (KB) of the cognitive nodes, will be used to update the weight of the heuristics during network operation.

We evaluate and compare the performance of the aforementioned algorithms using MATLAB® simulations. In the following, considered simulations' setup and performance metrics are discussed. Simulation results and a detailed analysis of the results are also presented in this section.

2.5.1 Simulation Setup

The network is set up as described in Figure 2.7, with randomly deployed sensor nodes, and fixed deployment of relay and cognitive nodes. Simulation parameters are as described in Table 2.6. Energy deductions at the local cognitive nodes (LCNs) and relay nodes (RNs) during data transmission are as represented in Table 2.7, based on the transmit powers. The transmit power at RNs is fixed at 3 dBm, and it can be adapted at the LCNs to improve the probability of successful transmission as described in [21]. Data delivery paths from source LCNs in the network are initially established based on AHP analysis of paths along next hop neighboring RNs. Heuristic learning is introduced in this simulation to increase the average success rate of data delivery to the sink, irrespective of the randomness with which the requests for different traffic types are generated in the network.

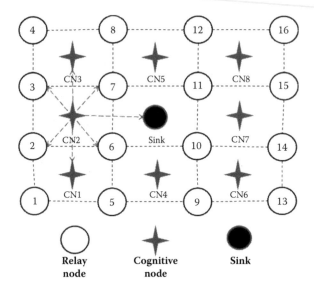

Figure 2.7   Use-case scenario setup for simulations.

Following are the three performance evaluation metrics that will be used to compare the performance of the aforementioned three algorithms:

1. **Network Lifetime:** Number of transmission rounds till all one-hop nodes to GCN/Sink node are dead (including RNs and LCNs).

Table 2.6   Simulation Parameters

PARAMETER	VALUE
Targeted area	1050 m × 1050 m
Number of nodes	SNs: 1500
	RNs: 16
	LCNs: 8
Transmit power	SN: <3dB
	RN: 3dB
	LCN: {3 dB, 5 dB, 7 dB}
Communication range	SN: 175 m
	RN: 250 m
	LCN: 350 m
	GCN: 500 m
Application payload size	121 Bytes
Per node offered load	0–1400 bits per second

Table 2.7 Transmit Power Consumption

PTX (dBm)	LIFECYCLE REDUCTION (UNITS)
3–5	2
5–7	3
7–9	4
≥10	5
3–5	2
5–7	3

2. **Success Rate (ρ):** It is defined as the ratio of the number of successful transmissions s to the sink over the total number of transmission rounds T during the network's lifetime. This is represented by Equation 2.1 as follows:

$$\rho = \frac{s}{T} * 100 \tag{2.1}$$

3. **Failure Rate (φ):** It is the ratio of the number of failed transmissions f to the sink over the total number of transmission rounds T during the network's lifetime. This is represented by Equation 2.2 as follows:

$$\phi = \frac{f}{T} * 100 \tag{2.2}$$

4. **eQoI:** *Effective-QoI* or eQoI is the heuristics estimate of the QoI associated with data delivered to the sink at the end of a successful transmission round. In other words, a heuristic estimate of the value of the QoI at the last hop that delivered the information to the sink.

2.5.2 Simulation Results and Analysis

Simulation results for the aforementioned three techniques are summarized in Table 2.8.

As shown in the table, results from the simulation using AHP analysis (AHPDD) suggests that during an average lifetime of 78 transmission rounds, the average success rate is 63 percent, and the average failure rate is 37 percent. However, during the worst case,

Table 2.8   Summary of Simulation Results

METHOD	LIFETIME (ROUNDS)	AVERAGE SUCCESS RATE	AVERAGE FAILURE RATE	BEST-CASE SUCCESS RATE	WORST-CASE FAILURE RATE
AHPDD	78	63	37	79	52
RL	106	47	53	68	63
CHAL	59	84	16	90	19
LDDA*	56	88	12	92	22

transmissions can fail for over 50 percent of the requests, as suggested by the worst-case failure rate.

With the cumulative-heuristic accelerated learning (CHAL), it was found that the average success rate increased to 84 percent, and the worst-case failure rate was as low as 19 percent. The best case success rate was 90 percent, which was only 6 percent off from the average success rate. This shows that the heuristics performed consistently well under various traffic loads and request arrival patterns. The performance of CHAL was matched very closely by the LDDA* heuristic search algorithm, which provided an 88-percent data delivery success rate, but a slightly higher failure rate of 22 percent in the worst-case scenario when compared with CHAL. Since it is more desirable to have a higher success rate in smart IoT applications, we further compare the performance of LDDA* and CHAL techniques in terms of their effective QoI (eQoI) as observed at the sink to identify the best heuristic of the two. Figure 2.8 shows the result of the comparison of the eQoI values for LDDA* and CHAL with AHPDD, which doesn't use any form of learning at the LCNs. In general, we observe that using some form of learning at the LCNs improves the eQoI of the data delivered to the sink. Of the learning techniques, we observe that LDDA* performs the best in terms of consistently delivering data with higher eQoI at the sink, even towards the end of the network's lifetime. Now, this eQoI is the hop-over-hop value of QoI associated with the data delivered to the sink with respect to latency, reliability and throughput. Apart from the hop-over-hop latency, the cumulative delay in receiving a response from the network for a request is reflected by the number of hops taken along the path from source to sink.

Thus, we can conclude that of the two proposed techniques, LDDA* is capable of delivering data to the sink with a higher average success rate, and better eQoI. Either of these techniques may be used

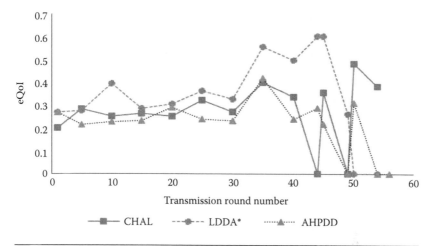

Figure 2.8   Comparison of eQoI as observed at the sink over the network lifetime.

for data delivery in the Cognitive ICSN for IoT applications, based on the application and end user requirements (i.e., eQoI), rate of successful data delivery, and cumulative delay from source to sink.

2.6 Conclusion and Open Issues

In this chapter, we have approached WSNs as information-centric networks, which enable retrieving more information than just the data in attribute–value pairs from the network. We reviewed some routing protocols in DCSNs and presented our views on how learning could have been used to improve their performance. Then we went on to see the various learning techniques available in the AI domain, and found that RL methods are more suitable for use with sensor networks. This is because reinforcement learning involves learning while interacting with the environment which is important in WSN environments where network dynamics change due to changing channel conditions, node deaths and changing traffic conditions, to name a few. We explored some more RL techniques, and analyzed what the ICSNs should learn, before arriving at a suitable technique for implementing learning in WSNs. Network Lifetime and Quality of Information are the primary features that must be improved by the use of learning algorithms. This work presents a preliminary assessment of the potential advantages of introducing learning in WSNs. A detailed assessment of the best way to do so, and comparison with other techniques with respect to their impact on ICSN lifetime and QoI awareness will provide a more accurate evaluation of the benefits of introducing learning in sensor networks. Thus, we open up the ICSN paradigm as a research area with interesting possibilities.

Acknowledgment

The author would like to thank his outstanding PhD student, Dr. Gayathri Singh, for her appreciated assistance in achieving this work.

References

1. D. H. Friend, "Cognitive Networks: Foundation to Applications," PhD Dissertation, Electrical and Comput. Eng., Virginia Polytechnic and State University, Blacksburg, VA, Mar. 6, 2009.
2. C. Bisdikian, L. M. Kaplan, and M. B. Srivastava, "On the Quality of Information in Sensor Networks," *ACM Trans. Sensor Network*, vol. 9, no. 4, article 48, July 2013.
3. D. J. Pearson and J. E. Laird, "Incremental Learning of Procedural Planning Knowledge in Challenging Environments," *Computational Intelligence*, vol. 21, no. 4, 2005.
4. M. T. Lazarescu, "Design of a WSN platform for long-term environmental monitoring for IoT applications," *IEEE J. Emerging and Selected Topics in Circuits and Systems*, vol. 3, no. 1, pp. 45–54, Mar. 2013.
5. ITU-T technology watch briefing report series, no. 4, "Ubiquitous Sensor Networks," Feb. 2008.
6. ITU-T, Series Y Recommendations: "Requirements for Support Ubiquitous Sensor Network (USN) applications and services in NGN environment," Y.2221, Jan. 2010.
7. K. Kalevi, "Quality of Experience in Communications Ecosystem," *J. UCS* 14, no. 5, pp. 615–624, 2008.
8. R. Hoes, T. Basten, C. K. Tham, M. Geilen, and H. Corporaal, "Quality-of-Service Trade-Off Analysis for Wireless Sensor Networks." *Performance Evaluation* 66, no. 3, pp. 191–208, 2009.
9. D. Chen and P. K. Varshney, "QoS Support in Wireless Sensor Networks: A Survey." in *International Conference on Wireless Networks*, vol. 233, pp. 1–7, 2004.
10. J. F. Martínez, A. B. Garcí, I. Corredor, L. López, V. Hernández, and A. Dasilva, "QoS in Wireless Sensor Networks: Survey and Approach." in *Proceedings of the 2007 Euro American Conference on Telematics and Information Systems*, p. 20, ACM, 2007.
11. V. Sachidananda, A. Khelil, and N. Suri, "Quality of Information in Wireless Sensor Networks: A Survey." *ICIQ (to appear)*, 2010.

12. C. Bisdikian, L. M. Kaplan, and M. B. Srivastava, "On the Quality and Value of Information in Sensor Networks," *ACM Transactions on Sensor Networks (TOSN)*, vol. 9, no. 4, pp. 1–26, 2013.

13. A. Tolstikov, C. K. Tham, and J. Biswas, "Quality of Information Assurance Using Phenomena-Aware Resource Management in Sensor Networks." in *Networks, 2006. ICON'06. 14th IEEE International Conference on*, vol. 1, pp. 1–7, 2006.

14. B. Ahlgren, C. Dannewitz, C. Imbrenda, D. Kutscher, B. Ohlman, "A Survey of Information-Centric Networking," *IEEE Communications Magazine*, vol. 50, no. 7, pp. 26–36, July 2012.

15. G. Anastasi, M. Conti, M. Di Francesco, and A. Passarella, "Energy Conservation in Wireless Sensor Networks: A Survey." *Ad Hoc Networks*, vol. 7, no. 3, pp. 537–568, 2009.

16. C. Intanagonwiwat, R. Govindan, D. Estrin, J. Heidemann, and F. Silva, "Directed Diffusion for Wireless Sensor Networking," *IEEE/ACM Transactions on Networking*, vol. 11, no. 1, pp. 2–16, 2003.

17. W. Heinzelman, J. Kulik, and H. Balakrishnan, "Adaptive Protocols for Information Dissemination in Wireless Sensor Networks," Proceedings of the 5th Annual ACM/IEEE International Conference on Mobile Computing and Networking (MobiCom_99), Seattle, WA, August 1999.

18. Y. Yu, D. Estrin, and R. Govindan, "Geographical and Energy Aware Routing: A Recursive Data Dissemination Protocol for Wireless Sensor Networks," UCLA Computer Science Department Technical Report, UCLA-CSD TR-01-0023, May 2001.

19. ZigBee Specifications [online]. Available at http://www.zigbee.org ZigBee Document 053474r17, Jan. 2008.

20. A. Al-Fagih, F. Al-Turjman, and H. Hassanein, "Ubiquitous Robust Data Delivery for Integrated RSNs in IoT," in IEEE Global Commun. Conf. (GLOBECOM'12), Anaheim, California, Dec. 3–7, 2012, pp. 298–303.

21. G. T. Singh and F. M. Al-Turjman, "A Data Delivery Framework for Cognitive Information-Centric Sensor Networks in Smart Outdoor Monitoring," Elsevier Computer Communications, Oct. 2013. [Unpublished.]

22. G. T. Singh and F. M. Al-Turjman, "Towards Prolonged Lifetime for Large-Scale Information-Centric Sensor Networks," QBSC 2014.

23. G. T. Singh and F. M. Al-Turjman, "Cognitive Routing for Information-Centric Sensor Networks in Smart Cities," IWCMC 2014.

24. A. Eriksson, B. Ohlman, and K. A. Persson, "What Are the Services of an Information-Centric Network, and Who Provides Them?" IEEE AP2PS, 2012.

25. F. Al-Turjman, A. Al-Fagih, and H. Hassanein, "A Value-Based Cache Replacement Approach for Information-Centric Networks," in *Proc. of the IEEE Local Computer Networks (LCN)*, Sydney, Australia, 2013, pp. 902–909.

26. F. Al-Turjman and H. Hassanein, "Enhanced Data Delivery Framework for Dynamic Information-Centric Networks (ICNs)," in *Proc. of the*

IEEE Local Computer Networks (LCN), Sydney, Australia, 2013, pp. 831–838.

27. A. Al-Fagih, F. Al-Turjman, W. Alsalih, and H. Hassanein, "A Priced Public Sensing Framework for Heterogeneous IoT Architectures," *IEEE Transactions on Emerging Topics in Computing*, vol. 1, no. 1, pp. 133–147, June 2013.

28. S. Haykin, "Cognitive Radio: Brain-Empowered Wireless Communications," *IEEE J Sel Area Comm*, no. 23, pp. 201–220, 2005.

29. J. Mitola and G. Q. Maguire, "Cognitive Radio: Making Software Radios More Personal," *IEEE Personal Communications*, vol. 6, no. 4, pp. 13–18, 1999.

30. R. W. Thomas, D. H. Friend, L. A. DaSilva, and A. B. MacKenzie, "Cognitive Networks: Adaptation and Learning to Achieve End-to-End Performance Objectives," *IEEE Commun. Mag.*, vol. 44, no. 12, pp. 51–57, 2006.

31. D. H. Friend, R. W. Thomas, A. B. MacKenzie, and L. A. DaSilva, "Distributed Learning and Reasoning in Cognitive Networks: Methods and Design Decisions," in *Cognitive Networks—Towards Self-Aware Networks* (Q. H. Mahmoud, ed.), pp. 223–246, John Wiley & Sons, Hoboken, NJ, 2007.

32. R. V. Kulkarni, A. Forster, and G. K. Venayagamoorthy, "Computational Intelligence in Wireless Sensor Networks: A Survey." *Communications Surveys & Tutorials, IEEE* vol. 13, no. 1, pp. 68–96, 2011.

33. A. Förster, "Machine Learning Techniques Applied to Wireless Ad-Hoc Networks: Guide and Survey," in *Proc. 3rd Intl. Conf. Intelligent Sensors, Sensor Networks and Information Processing* (ISSNIP), 2007.

34. A. Förster and A. L. Murphy, "FROMS: Feedback Routing for Optimizing Multiple Sinks in WSN with Reinforcement Learning," in *Proc. 3rd Int. Conf. Intelligent Sensors, Sensor Netw. Inf. Process. (ISSNIP)*, 2007.

35. A. Förster and A. L. Murphy, "Balancing Energy Expenditure in WSNs through Reinforcement Learning: A Study," in *Proc. of the 1st Int. Workshop on Energy in Wireless Sensor Networks (WEWSN)*, 2008.

36. M. Di and E. Joo, "A Survey of Machine Learning in Wireless Sensor Networks," in *Proc. 6th Int. Conf. Inf., Commun. Signal Process*, 2007.

37. R. S. Sutton and A. G. Barto, Reinforcement Learning: An Introduction, MIT Press, Cambridge, MA, 1998.

38. S. Kaplantzis, A. Shilton, N. Mani, Y. A. Sekercioglu, "Detecting Selective Forwarding Attacks in Wireless Sensor Networks using Support Vector Machines," *Intelligent Sensors, Sensor Networks and Information, 2007. ISSNIP 2007. 3rd International Conference on*, 335–340, Dec. 3–6, 2007.

39. P. K. Biswas, "Architecting Multi-Agent Systems with Distributed Sensor Networks," *Integration of Knowledge Intensive Multi-Agent Systems, 2005. International Conference on*, pp. 161–166, April 18–21, 2005.

40 K. Shenai and S. Mukhopadhyay, "Cognitive Sensor Networks," in *Proc. IEEE 26th Int. Conf. Microelectronics (MIEL)*, May 2008, pp. 315–320.

41. P. Boonma and J. Suzuki, "Exploring Self-Star Properties in Cognitive Sensor Networking," in *Proc. IEEE/SCS Int. Symp. Performance Evaluation Comput. Telecommun. Syst. (SPECTS)*, Edinburgh, UK, pp. 36–43, Jun. 2008.

42. C. Perera, A. Zaslavsky, P. Christen, and D. Georgakopoulos, "Context Aware Computing for the Internet of Things: A Survey," *IEEE Commun. Surveys Tuts.*, vol. 16, no. 1, pp. 414–454, 2014.

43. W. S. Terry, *Learning and Memory: Basic principles, Processes, and Procedures*, Pearson Education, Inc., Boston, 2006.

44. C. G. Atkeson and J. C. Santamaria, "A Comparison of Direct and Model-Based Reinforcement Learning," *International Conference on Robotics and Automation*, 1997.

45. R. A. C. Bianchi, C. H. C. Ribeiro, and A. H. R. Costa, "Heuristic Selection of Actions in Multiagent Reinforcement Learning," *IJCAI'07*, Hyderabad, India, 2007.

46. R. A. C. Bianchi, C. H. C. Ribeiro, and A. H. R. Costa, "Heuristically Accelerated Reinforcement Learning: Theoretical and Experimental Results," in *ECAI*, pp. 169–174, 2012.

47. G. Gigerenzer and W. Gaissmaier, Heuristic Decision Making. *Annual Review of Psychology* 62, pp. 451–482, 2011.

48. G. Gigerenzer, P. M. Todd, and the ABC Research Group, *Simple Heuristics that Make Us Smart*, Oxford University Press, New York, 1999.

49. R. A. C. Bianchi, R. Ros, and R. L. de Mantaras, "Improving Reinforcement Learning by Using Case Based Heuristics," in *Case-Based Reasoning Research and Development*, pp. 75–89. Springer, Berlin–Heidelberg, 2009.

50. H. Geffner, "Heuristics, Planning and Cognition," in *Heuristics, Probability and Causality: A Tribute to Judea Pearl* (eds. R. Dechter, H. Geffner, and J. Halpern), College Publications, London, 2010.

51. A. Tversky and D. Kahneman, "Judgment under Uncertainty: Heuristics and Biases," *Science*, New Series, vol. 185, no. 4157, pp. 1124–1131, 1974.

52. R. Davis, H. Shrobe, and P. Szolovits, "What Is a Knowledge Representation?" *AI Magazine*, vol. 14, no. 1, pp. 17–33, 1993.

3

COGNITIVE-NODE ARCHITECTURE AND A DEPLOYMENT STRATEGY FOR THE FUTURE SENSOR NETWORKS*

3.1 Introduction

Information-centric sensor networks (ICSNs) are a class of context-aware communication networks that provide an infrastructure for knowledge-based intelligent information service to anyone, anywhere, and at any time [1]. IEEE 802.15.4/ZigBee-based wireless sensor networks (WSNs) provide the basic infrastructure to deliver sensed information to end users in diverse application environments such as agricultural monitoring in rural areas, structural health monitoring of buildings and bridges in urban areas, tracking items in industrial supply chain management applications, detection of forest fires, and even landmines detection in former war zones [2]. These ICSN applications that require a large-scale deployment of the sensor network in order to cover the large target areas and provide more sensing points in the region being monitored, the network topology changes dynamically due to node deaths, changing node associations and varying environment conditions, thus affecting the network connectivity [3] and information gathering and delivery capabilities. In addition, the network may have to deal with service requests coming from a variety of end users, including individual consumers, public enterprises, government organizations, and even machines that

---

* This article was originally published in *Mobile Networks and Applications*. F. Al-Turjman, Cognitive node architecture and a deployment strategy for the future WSNs, pp. 1–19, 2017. Reprinted with permission.

are information-monitoring devices. It is very challenging for sensor networks in their current form to provide a common platform that can support such diverse ICSN applications, while providing context-aware information to end users that differ in their requirements on the attributes associated with service-data such as reliability, latency, and throughput. As in the case with Internet of things (IoT) applications, WSNs are not equipped to handle the heterogeneous traffic, nor do they have adequate capacity to store the large volume of data generated as a result of the multiple requests being serviced by the network and need modifications in the infrastructure to support the functionality [4].

To improve the capabilities of the network that delivers data to IoT environment, we propose the use of cognitive nodes (CNs) based on the elements of learning, reasoning, and knowledge representation in the underlying network. CNs will provide enhanced capabilities to the WSN to deal with the network connectivity and node dynamics in large-scale deployments. They will also provide space for local storage of data before data gets delivered to the end user. This will help to maintain their availability at intermediate locations, other than their points of publication (i.e., the sensor nodes) for ease of access. To this end, the main contributions of this chapter are as listed below:

- We provide a description of the conceptual architecture of the cognitive node and the components that constitute its cognitive elements, that is, knowledge representation, learning, and reasoning; and describe their functions.
- We identify a grid-based deployment strategy for relay and cognitive nodes in the large-scale WSN such that the probability of successful data reception between the communicating nodes is greater than 0.8.
- We also calculate the number of relay and cognitive nodes required to cover the target area while ensuring a high probability of successful data reception.

The remaining sections have been organized as follows: In Section 3.2 we present the related work, the conceptual architecture of the cognitive node, and a brief description of the utilized cognitive elements. Section 3.3 discusses our system models and targeted problem statement. The cognitive deployment strategy and utilized cognitive elements is presented in detail in Section 3.4. Next, extensive simulation results are presented in Section 3.5 before concluding the chapter in Section 3.6.

3.2 Related Work

Sensor node deployment problem has been extensively studied in literature over the past decade. Researchers have considered various factors such as coverage, connectivity, energy-efficiency, and fault-tolerance while proposing deployment strategies for sensor nodes (SNs) [5]. With the introduction of the ZigBee standard [6], the focus shifted from sensor node to relay node (RN) placement problem, as the RNs could serve to maintain connectivity of sensor nodes with their base station even when the network size scaled-up [7]. The RNs increased the communication range of SNs and also took over the energy demanding task of data communication within the network from the SNs. This in turn increased the lifetime of the SNs, thus improving the longevity of the network. However, as WSN applications evolved from simple event monitoring or tracking applications to complex applications such as ecological monitoring [8], network deployment and its operational complexity increased. The WSN had to not only provide periodically monitored data, but had to even respond to on-demand queries and emergency situations. The changing application requirements made the network traffic very heterogeneous, leading to load balancing issues among the nodes and traffic bottlenecks in the network. Recent research has even considered the use of mobile data collectors, traffic-aware relay node deployment and artificial intelligent (AI) techniques to manage the dynamic network [9]. But data latency and reliability become an issue when mobile data collectors are used, and AI techniques have targeted very specific applications [10]. They have not been architecturally developed and implemented in a way that can be extended to different WSN application platforms. Thus we say that in their current state, WSNs with SNs, RNs and data collector nodes will not be able to understand and respond to changing application requirements. The network will not be able to cater to performance attributes of latency, reliability, energy consumption, and fault-tolerance while delivering data to the sink. We collectively call these attributes the quality of information (QoI) attributes [11], as they represent the attributes that the application layer would associate with the data delivered to the sink, to measure the application-awareness of the response generated by the network to the end user's request. In order to make the network aware of the changing application requirements, and enable it to provide QoI aware data, we propose the use of special nodes called *cognitive nodes* (CNs) in the underlying WSN. These CNs when strategically deployed in the network will ensure data-delivery with user-desired QoI to the sink, in each round of data transmission throughout the lifetime of the network. We will refer to this network as an *information–centric sensor network* (ICSN) from this point forward, as it will draw on the features of information-centric networks (ICNs) in terms of named-data association, in-network caching and the use of CNs as intermediate nodes that will process and store the information within the network [12,13]. However, we must mention that the idea of named-data association in WSNs is not new. The idea has existed in data-centric sensor networks (DCSNs), which are a special class of WSNs that function as information-retrieval networks rather than serving as point-to-point communication networks [2,14]. Sensor attributes are used for data gathering and delivery, which makes the use of node addresses inessential. This can lead to huge energy savings for the sensor network, as a single query can be broadcast throughout the network to gather all relevant data from different sources, as against multiple queries addressed to specific locations to gather the same data. This translates to energy savings for all the network nodes, leading to prolonged network lifetime.

Shifting our focus back to the cognitive nodes, the information-centric approach to query dissemination used by these nodes helps in finding only relevant data and changes the way the network handles user requests. There is awareness in the network, about the specific information requested by the user; that is, temperature data or humidity information from a specific geographic area, at a specific time in the present, or from sometime in the past. In addition, the CNs enable the network to understand the QoI with which it is expected to return the requested data to the sink and the network is able to adapt the use of its resources to find paths that are either reliable, have low latency or offer a high throughput. This way, the network is not always exerting itself to find the best path that satisfies all the attributes, but prioritizes the QoI attributes for each transmission round based on the end user requirements, and finds a suitable path accordingly, thus prolonging the network lifetime. Now the challenge is in finding the best place to deploy these nodes in the ICSN. Optimal node placement is a very challenging problem and has been proven to be NP-hard [5]. With CNs, there are constraints on how many such nodes can be used in the network, and whether one can use only CNs or combine it with the use of RNs in the underlying network.

In this chapter, we identify the cognitive functions of the CN, address its deployment problem, and, through simulations, identify the best combination of RNs and CNs that the network can benefit from to minimize energy consumption and prolong network lifetime [15] while catering to the QoI attributes of reliability and instantaneous throughput. Thus contributing to not only good quality of user experience, but also improving the lifetime of the network during which data is delivered to the end user–based on user-desired QoI attributes.

Cognition is the process by which knowledge is acquired through intuition, perception, planning, and reasoning. If a communication network is capable of observing the impact of its own actions on the environment, and learns to use the knowledge acquired from accumulating these observations to adapt the data delivery paths according to user requirements and changing network conditions, then it is said to be exhibiting cognitive behavior [16]. From this perspective, we identify knowledge representation, learning, and the ability to infer from the knowledge acquired, as elements of cognition sufficient to achieve our proposed goals. Figure 3.1 represents the three major components that we define for our cognitive nodes: learning, reasoning, and knowledge representation; and associates them with their respective functions. These elements of cognition, when incorporated in the network nodes of a WSN, help it in better understanding and catering to the end user requirements or what we call quality of information (QoI) requirements.

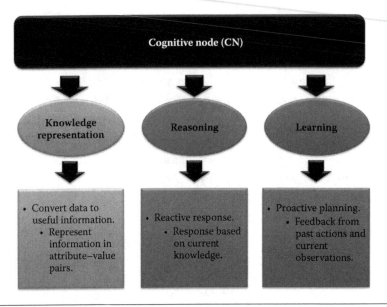

Figure 3.1   The elements of a cognitive node.

Our expectation from the cognitive elements in Figure 3.1 would be mainly to cater for the following objectives:

- Reactive response: Observe current network behavior and respond adaptively to changing network dynamics
- Proactive planning: Learn from past behavior and plan for the future so as to make predictions and decisions that positively impact the network survivability and application QoI during its lifetime.

Using these elements, a conceptual architecture of the cognitive node is illustrated as depicted in Figure 3.2. The ultimate goal to be achieved by CNs is to reduce the periodic data transmission in favor of a planned data collection mechanism that is optimized to match the end user's requirements while meeting the coverage, connectivity, and longevity requirements of the sensor network. In order to achieve that, the CNs will need to store information about the network status, as well as have a learning mechanism that enables decision making based on past experiences. End user requirements will be received via a CN's transceiver, while environment observations are to be collected via the sensor units. The two sources of information are consolidated in a knowledge base, which will provide this information to a learning

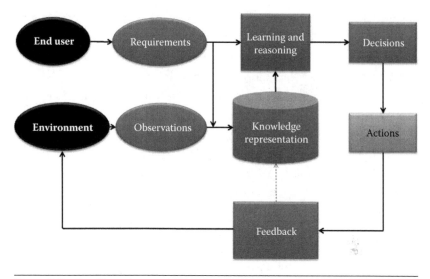

Figure 3.2 Cognitive node conceptual architecture.

engine. The learning engine analyzes observations and requirements in order to form decisions as to how the requirements will be satisfied given the network's current status and its past response to similar situations from the knowledge base. The outcome is a set of actions that comprise the final plan that will serve a specific request.

3.3 System Models

In this section, we describe the assumed network, communication, cost, and energy models of the targeted cognitive ICSN and generalize a deployment problem statement in this domain.

3.3.1 Network Model

Nodes in WSNs can be deployed in a flat, hierarchical, or geographic location based strategy. In terms of energy conservation, hierarchical deployment strategies provide better performance for WSNs [2]. We had proposed a hierarchical strategy for cognitive communication in WSNs in our earlier work [17,18]. We make use of this hierarchical approach here for a large-scale deployment in IoT applications. Cognitive nodes (CNs), relay nodes (RNs), sensor nodes (SNs), and a sink node are the node-level entities of the network. CNs act as cluster heads for RNs and are the decision makers for the network. Using Radial graph [19], distributed RNs use the aforementioned QoI attributes to break local symmetries between the RNs and distinguish those who are more capable to be RNs cluster heads (i.e., CNs). In other words, each node is made a cluster head as long as it has locally the largest attribute values. Once a node is dominated by a cluster head, it abstains from local competition, giving other nodes the same chance in the future. We let this duty rotates among various members of the WSN via periodic reelection for CNs. This is very useful when energy reserves are used as discriminating attributes.

In the proposed network model, CNs process the requests received from the end user by making use of the cognitive elements of knowledge representation, learning, and reasoning to identify a data delivery path to the sink that meets the user's quality of information (QoI) requirements. RNs act as cluster heads for SNs and also participate in relaying information from CNs to the sink. Sensor nodes gather sensed data and forward it to both RNs and CNs lying within their communication range. The hierarchical deployment strategy helps to distribute the tasks between the relay and cognitive nodes and better manage the network connectivity. We assume that RNs and CNs have the same communication range for a given transmit power. However, the transmit power at RNs is fixed (0 dB) at the time of deployment and CNs are allowed to adapt their transmission power from a predetermined set of values (–3 dB to + 10 dB) to achieve the desired transmission range and QoI.

3.3.2 Energy Consumption Model

The voltage discharge characteristics of most lithium AA batteries (irrespective of their chemistry) suggests that once the terminal voltage drops to about 30 percent of its original value, almost all of the battery's usable energy is depleted. Lithium batteries typically last 500–1000 cycles before the terminal voltage drops to this value, depending on the application and environment in which it is operated [20]. In our system, we assume that the batteries at RNs and CNs are capable of delivering consistent performance for about 500 cycles, after which they are assumed to be drained out of energy. Rechargeable batteries, or batteries of higher energy ratings, may be considered for use at CNs when compared to RNs, as their processors may consume additional energy during information processing, adaptive transmission, and information feedback for the cognitive decision process. So, the actual cost of RNs and CNs may not be very different, but the energy consumption of CNs could be higher due to the following reasons: (1) CN's ability to adapt transmit power to alter their communication range during the data transmission phase, and (2) CN's processors operate for longer (when the nodes are active, but not transmitting data), to perform additional computing and information processing required during the cognitive decision making process. If energy consumption at a RN is denoted by E_{RN}, and that at CN is denoted by E_{CN}, then following the energy consumption discussion above, we can say that energy consumption at cognitive node is greater than that at relay nodes, as depicted by Equation 3.1.

$$E_{CN} > E_{RN} \tag{3.1}$$

We set the initial battery cycle life to 500 units and every time a node is involved in a data or control message communication, we reduce the node's battery cycle life as shown in Table 3.1, based on the transmit power used for communication. And thus, at the end of every round, the total energy consumed per node i can be written as

$$E^i_{cons} = \sum_{\text{per round} J_{tr}} + \sum_{\text{per round} J_{rec-}}, \tag{3.2}$$

where $J_{tr} = L(\varepsilon_1 + \varepsilon_2 d^n)$ is the energy consumed for transmitting a data packet of length L to a receiver located d meters from the transmitter. Similarly, $J_{rc} = L\beta$ is the energy consumed for receiving a packet of the same length [15,21]. In addition, ε_1, ε_1, and β are hardware specific parameters of the used transceivers. In addition, if the initial energy E_{init} of each node is known, the remaining energy per node i at the end of each round is

$$E^i_{rem} = E^i_{init} - E^i_{cons}, \tag{3.3}$$

Table 3.1 Reduction in Cycle Life Based on Transmit Power

P_t	CYCLE LIFE REDUCTION (UNITS)
<3 dBm	1
3 dBm–5 dBm	2
5 dBm– 7 dBm	3

3.3.3 Communication Model

In practice, the signal level at distance r from a transmitter varies depending on the surrounding environment. These variations are captured through the so-called log–normal shadowing model. According to this model, the signal level at distance r from a transmitter follows a log–normal distribution centered on the average power value at that point [22]. Mathematically, this can be written as

$$P_{recv}(d) = P_t - P_L(d) = P_t - P_L(r_0) - 10n \log\left(\frac{r}{r_0}\right) + \chi \qquad (3.4)$$

where $P_{recv}(d)$ is the received signal power at distance d, P_t is the source node transmission power, $P_L(r_0)$ is the path loss measured at reference distance r_0 from the transmitter, n is an environment dependent path loss exponent, and χ is a normally distributed random variable with zero mean and variance σ^2, i.e. $\chi \sim \mathcal{N}(0,\sigma^2)$. With the aid of this model, the probability of successful communication between two nodes separated with a distance r can be calculated as follows. Assume P_{min} is the minimum acceptable signal level for successful communication between a source S and a destination D separated by distance r. The probability of successful communication is $\rho[S, D] = Pr[P_{recv}(r) \geq P_{min}]$. After some mathematical manipulations, $\rho[S, D]$ can be written as

$$\rho[S,D] = Q\left(\frac{P_{min} - P_t - P_L(r_0) - 10n \log\left(\dfrac{r}{r_0}\right)}{\sigma}\right) \qquad (3.5)$$

where $Q(\cdot)$ is the Q-function defined as

$$Q(x) = \frac{1}{\sqrt{2\pi}} \int_x^\infty e^{-t^2/2} dt \qquad (3.6)$$

In this work, the probability of successful communication between nodes i and j should exceed a certain threshold, γ_{th}. Hence, the condition $\rho[i, j] \geq \gamma_{th}$ will be used. The value of Pr can be estimated using a cumulative density function as follows:

$$Pr[P_{recv}(d) > \gamma_{th}] = Q\left[\frac{\gamma_{th} - P_{recv}(d)}{\sigma}\right] \qquad (3.7)$$

3.3.4 Operational Cost Model

In a WSN deployment consisting of sensor, relay, sink, and cognitive nodes, the sensors used for application-relevant data gathering are typically the most expensive hardware component. Sensors are deployed on sensor nodes which are powered by single-use batteries which are typically not replaceable due to accessibility issues. They operate at low transmit powers (typically less than 3 dB) and have a relatively small communication range (a few hundred meters) when compared with RNs and CNs. To have the sensors remain operational for the longest duration possible, it is best to let them operate in sleep mode more often, and turn them on only when data needs to be gathered from their surroundings. Comparing relay and cognitive nodes, their hardware costs are very close in terms of the batteries and processors used. However, they differ in their energy consumption model, as mentioned above.

In addition, CNs may incur a slightly higher hardware cost in terms of having an additional flash memory storage, where it can store the data gathered from nearby sensors and relay nodes, and also the information observed from the network interactions. This way, CNs can better perform information caching functions in the information-centric sensor network environment. If we represent the hardware cost of RNs as C_{RN-HW}, and that of CNs as C_{CN-HW}, then following the above discussion, we can arrive at Equation 3.8, which suggests that the cost of CN's hardware is equal to or greater than the cost of the relay node hardware, based on the use of the additional flash storage.

$$C_{CN-HW} \geq C_{RN-HW} \qquad (3.8)$$

Thus, comparing the energy consumption model and hardware costs of the RNs and CNs from Equation 3.1 and Equation 3.8, we can say that the cost of operating CNs, or its operational cost (OC_{CN}) is more than the operational cost of RNs (OC_{RN}), as shown in Equation 3.9.

$$OC_{CN} > OC_{RN} \qquad (3.9)$$

Hence, it is important to consider reducing the number of cognitive nodes used in the deployment strategy in order to reduce the operational cost of network and its maintenance thereafter.

3.3.5 Problem Definition

For large-scale WSNs, we define the node deployment problem as follows: "Determine the number and location for the placement of relay and cognitive nodes in a given targeted area such that (1) the probability that the received signal strength is above a specific threshold γ_{th} that guarantees the QoI, (2) the network is connected in such a way that there is a path from each SN to the sink through the RNs or CNs, and (3) the deployed network satisfies the three main QoI attributes: High instantaneous throughput, low average delay per node, and reliable data delivery over the multi-hop communication."

Per this definition, it's worth pointing out that the probability of the received signal strength being above a threshold value is defined as the probability of successful data reception Pr as described in Equation 3.7. As identified in the cost model, we want to minimize the number of CNs and keep their number lower than the number of RNs to minimize the total cost of the network. We also want to assure that there is at least one RN/CN for each sensor node to deliver its information so that SNs are only involved in short-range local communications that incur the minimum cost. In the following section we identify a strategy for the deployment of CNs and RNs for large-scale ICSN applications.

3.4 A Cognitive Deployment Strategy for ICSN

In order to determine the deployment strategy for the relay and cognitive nodes in the network, we make the following assumptions:

- Sensors nodes are deployed randomly but uniformly throughout the target area. They have a fixed transmit power with a communication range of 175 m.
- SNs can communicate and bind with RNs and CNs in a single-hop, but do not communicate with each other.
- RNs have a fixed transmit power and communication range, but the values are higher when compared with those of SNs.
- CNs can vary their transmit power to increase their transmission range to values higher than those of RNs. However,

when their transmit power is the same value as that of RNs, they offer the same communication range.

- We approximate the target area to be a square region and divide the entire area into smaller squares of side L.
- The sink is deployed at approximately the center of the target area.

Given these assumptions, the goal of the deployment strategy is to identify the length L of the side of each square grid cell [23] and the position of the RNs and CNs on the grid such that the RNs can communicate with at least one CN, and Pr > 0.8 along each hop of the data delivery path from a source node to the sink. We device such a node deployment strategy by making use of the following basic properties of a square.

Property 1. Center of a square of side L is equidistant from each of its vertices, and has a length $L/\sqrt{2}$.

For a square of side L, the diagonals intersect at the center of the square, and are perpendicular bisectors of each other. Using Pythagoras's theorem [24], we know that the length of the diagonal which forms the hypotenuse of the isosceles triangle formed by the two sides of the square, can be found as $\sqrt{(L^2 + L^2)} = \sqrt{2} * L$. Thus the center of a square is equidistant from each of its vertices and has a length $L/\sqrt{2}$, as shown in Figure 3.3(a).

Property 2. The distance from the center of the square to any of its vertices is the maximum separation distance that can be achieved between the center and any other point on the square.

If we draw a circle with radius $L/\sqrt{2}$, whose center lies at the center of a square of side L, we would be circumscribing a circle that passes through each of the vertices of the square. This circle does not touch any other point of the square other than the vertices. Thus the distance between the center of the square and its vertices is the maximum separation distance that can be achieved between the vertex and any other point within the square.

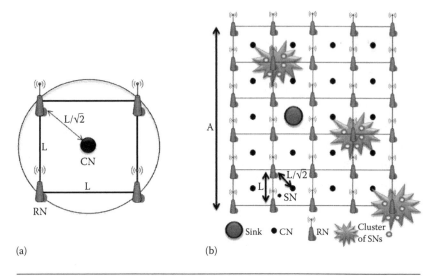

Figure 3.3 (a) Square grid cell of side L having relay nodes (RN) at the corners, and (b) RNs, CNs, and inquired clusters of SNs on the grid.

Now let us consider the path loss (PL) and received signal strength P_{recv}. From the log–normal shadowing path loss communication model and Equation 3.4, we know that the path loss increases as the separation distance between a transmitter and receiver node increases. Conversely, if the distance between two communicating nodes is reduced from $\sqrt{2} * L$ across the diagonals of a square of side L, to $L/\sqrt{2}$ at half the distance, the increase in received signal strength P_{recv} can be calculated from Equation 3.4 as follows:

$$P_{recv}(\sqrt{2} * L) - P_{recv}\left(\frac{L}{\sqrt{2}}\right) = PL\left(\frac{L}{\sqrt{2}}\right) - PL(\sqrt{2} * L) \quad (3.10)$$

Expressing $P_{recv}\left(\dfrac{L}{\sqrt{2}}\right)$ in terms of $P_{recv}(\sqrt{2} * L)$,

$$P_{recv}\left(\frac{L}{\sqrt{2}}\right) = P_{recv}(\sqrt{2} * L) + 1.5 * n \quad (3.11)$$

Thus the received signal strength P_{recv} at distance $\dfrac{L}{\sqrt{2}}$ is increased by an amount of 1.5*n, when compared with its value at double the distance $\sqrt{2} * L$. Using the result from Equation 3.11, we can say that for a two-dimensional square grid cell with transmitter nodes placed at the vertices, an intermediate node placed at the center of the square can improve the probability of reliable reception by 1.5*n, using a two-hop communication. Next, we describe the algorithm for deployment of RNs and CNs in the targeted area.

In Algorithm 3.1, lines 1 to 5 describe the inputs required to come up with the deployment plan. The size of the target area, number of SNs available and their communication range and the location of the sink are essential to decide on the outputs described in steps 7 and 8. They are the number and position of RNs and CNs required for maintaining connectivity of the SNs with the sink. In Step 9, the receiver sensitivity of the RNs and CNs is set to −101 dBm, which is typically the value in commercially available SNs and RNs. In step 10, a threshold value of signal strength γ^{th} is set such that it is 3 dBm above R_{sense}. This is to guarantee reception of signals that are stronger than the receiver's sensitivity, which is the least value of the signal that it can detect. Once these values are set, we use Equation 3.7 to plot a graph of the variation of Pr as a function of d at transmit powers in the range (−5 dBm to 10 dBm). The transmitter and receiver represent the RNs and CNs. In line 13, the values obtained from the plot are tabulated to strategically identify a value of d in step 14 to ensure that there is at least one RN or CN lying between any two SNs to guarantee connectivity of SNs with the sink across the entire network. In line 15, we ensure that the transmit power for the chosen d is able to support Pr > 0.8 for every link, at least under near-ideal conditions. Once the side of each square grid is identified in step 16, steps 17–21 describe the steps to identify the number of rows and columns in the square grid covering the target area and number and position of RNs and CNs in each grid cell. Thus, for SNs placed uniformly, randomly in a target region, Algorithm 3.1 gives the deployment plan for placing RNs and CNs in the area. From this plan, we arrive at L = 350*m*, Number of CNs = 9, Number of RNs = 16. Figure 3.3(b) illustrates the deployment of the relay and cognitive nodes in a square grid using this strategy. Summarizing this deployment strategy, we conclude the following:

> For a square target area of side A, and square grid of side L, we have $\sqrt{G}$ number of rows of square grids, G number of cognitive nodes and $(\sqrt{G}+1)^2$ number of relay nodes in the network, where G is A^2/L^2 rounded off to the nearest higher perfect square number. Table 3.2 shows the values of Pt and d for Pr > 0.9 and Pr > 0.8 are tabulated using values from the simulation results in Figure 3.4.

Table 3.2   Values of D for Different Transmit Powers for $P_R > 0.8$ and $P_R > 0.9$

	PR > 0.9		PR > 0.8
P_T (DBM)	D (M)		D (M)
−3	150		190
0	200		225
3	250		280
5	275		300
7	300		350
10	360		400

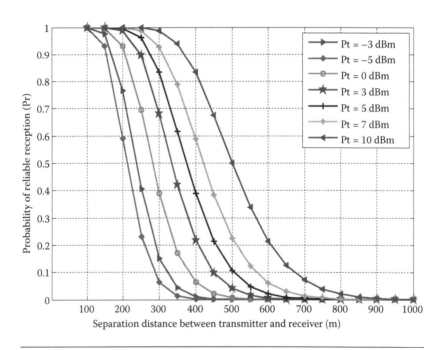

Figure 3.4   Plot of probability of received signal strength versus the separation distance 'd'.

Algorithm 3.1 Deployment Plan for RNs and CNs in Target Area

1. **Inputs:**
2. Target area A^2: 1050 m × 1050 m
3. Number of sensor nodes: 1500
4. Sensor node communication range (r_{SN}): 175 m
5. Sink position: Center of target area
6. **Outputs:**
7. Number of RNs and CNs for the target area
8. Position of RNs and CNs in the deployment region
9. **Begin:**
10. **Initialize:** Receiver sensitivity R_{sense}= −101 dBm
11. Threshold signal strength γ_{th} = −98 dBm
12. Plot a graph of Pr against d, for different transmit powers
13. Tabulate values from the plot in step 12
14. Identify a value of d such that $d \le 2^*(r_{SN})$ and $Pr > 0.8$
15. Choose Pt such that $0 \le Pt \le 10$, for $Pr > 0.8$ at d
16. Set d as the side L of each square grid in the target area A^2
17. Approximate number of square grids required to cover the target area x is A^2/L^2
18. Round off x to the nearest higher whole number G, such that G is the perfect square of a whole number
19. $\sqrt{G}$ is the number of rows and columns of square grids in the targeted area.
20. G will be the total number of CNs in the network, each deployed at the center of a square grid at a distance $L/\sqrt{2}$ from the corners of the grid
21. $(\sqrt{G}+1)^2$ will be the total number of RNs in the network, placed at the corners of each of the square grid
22. **End**

Thus, to cover a square target area of side A, unit square grid cells of side L can be replicated over the entire area to ensure connectivity among all the nodes. Following the steps in Algorithm 3.1, we know that G is not the least number of such square grid cells required to cover the target area, but a value chosen to retain the approximation of the target area as a square one, while maintaining the size of unit grid cells constant. We know that the separation between nodes at the corners and central node is maximized and the central nodes will be equidistant when juxtaposed with similar cells, due to the properties of the square. Thus, we can say that the number of RNs and CNs identified in Algorithm 3.1 provide a feasible deployment plan. Meanwhile, elements of cognition in the CNs form the two main constituents of our proposed deployment strategy. These elements are *reasoning* and *learning*.

3.4.1 Learning

Learning is used in our deployment plan in order to determine the most appropriate paths towards the sink that satisfy the QoI requirements. This cognition element uses a direction-based heuristic to determine RNs' locations that lie in the most appropriate direction of the sink. Hence, each time a CN has to choose the next hop, the direction-based heuristic eliminates RNs' positions that increase the distance between the current RN and sink node. Knowledge of the positions of the CN and its one-hop RNs is used by the heuristic to determine the set of such RNs, which we call forward-hop-RNs. Thus the forward-hop-RNs of a CN identified by the direction-heuristic is constituted by those RNs that reduce the distance between the CN and the network sink. This information is stored in the CN for use in the next transmission round. Thus the direction-based heuristic, along with feedback from the network about the chosen RNs' positions helps the CNs to learn candidate positions for the forward-hop-RNs to the sink, as the network topology changes.

Example 3.1: Assume S_1 and S_2 have data to be sent to destination nodes D_1 and D_2. R_n are all the available grid vertices towards the destination sink. Out of these positions, it is determined that R_5 as shown in Figure 3.5(a) has the lowest link outage probability to D_1 and D_2. Therefore, S_1 initiates a path to R_5. Meanwhile, S_2 also forward a high traffic of data to R_5. When multiple source nodes start exchanging their data with R_5 as well, the route to R_5 may get congested. A cognitive network with *learning* capabilities will be able to identify the congestion at R_5 (by observing the decrease in throughput). Sharing this observation with neighboring nodes, the corresponding CN would be able to respond to the congestion proactively,

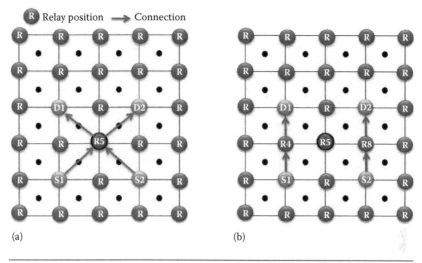

Figure 3.5   (a) Typical positioning and (b) Cognitive positioning in IoT-networks.

by replacing the utilized RN at R_5 with different set of candidate relays at R_4 and R_8 as shown in Figure 3.5(b).

3.4.2 Knowledge-Representation and Reasoning

In our cognitive approach, we assume a modified version of the analytic hierarchy process (AHP) [25] for implementing the reasoning element of cognition in the IoT. AHP supports multiple-criteria decision making while choosing the next hop position. For example, if we have delay-sensitive data, the node which provides the lowest latency will be chosen even though it might degrade other metrics such as the network energy or throughput. If two next hops guarantee the same latency then the next attribute to compare will be energy, and then, throughput, assuming that energy is the next desired attribute in the IoT-network. AHP provides a method for pairwise comparison of each of the attributes and helps to choose the node that can provide the best network performance on the long run. The following subsequent example has more details on the utilized AHP. While AHP calculations help in deciding the next hop, it also help in planning for future actions. The CNs are able to store the calculated values of the QoI-attributes, which can be used in future transmission rounds. Hence, these values are not necessarily calculated at every transmission round.

Example 3.2: Assume a three-level hierarchy in the AHP: *Goal (determine the next RN), Criteria (Reliability, Energy, and Throughput)* and *Alternatives (RN1, RN2, and RN3)*. A fundamental scale for pairwise comparisons is then used to set priorities for the QoI attributes/criteria at the CNs. Given the very limited energy constraint in WSNs, we would assign the highest priority to energy, followed by *reliability* and then *throughput*. We tabulate the relative priorities of these attributes using pairwise comparison and generate Table 3.3. From Table 3.4 we generate Table 3.5. Then, we apply the following steps:

1. Represent the values of Table 3.4 in the matrix form

$$A = \begin{bmatrix} 1 & 4 & 6 \\ 1/4 & 1 & 3 \\ 1/6 & 1/3 & 1 \end{bmatrix}.$$

2. Compute the Eigen vector of the matrix A.
3. Isolate the absolute, real values of the Eigen vector.
4. Compute the relative priority values as shown in Table 3.5 in order to decide which RN position to choose.

Table 3.3 Pairwise Comparison of the QoI-Attributes

Energy	4	Reliability	1
Energy	6	Throughput	1
Reliability	3	Throughput	1

Table 3.4 AHP for QoI Attributes vs. Goal

GOAL: BEST ATTRIBUTE	ENERGY	RELIABILITY	THROUGHPUT	RELATIVE PRIORITIES OF THE ATTRIBUTES
Energy	1	4	6	0.691
Reliability	1/4	1	3	0.2176
Throughput	1/6	1/3	1	0.0914

Table 3.5 AHP Evaluating the Overall Priorities for all Possible RNs

BEST CANDIDATE FOR NEXT-HOP RN$_X$	PRIORITY WITH RESPECT TO			
	ENERGY	RELIABILITY	THROUGHPUT	GOAL
RN$_1$	0.252	0.015	0.101	0.375
RN$_2$	0.2	0.018	0.11	0.329
RN$_3$	0.164	0.019	0.116	0.296

Due to large-scale applications targeted in this research, sensor nodes are assumed to be randomly distributed as we aforementioned. They form clusters based on their vicinity once the RNs are deployed. To do this, a hybrid energy efficient distributed (HEED) approach [26,27] for ad-hoc sensor networks is applied to assign a relay node as a Cluster Head (CH) for those clustered sensors (see Figure 3.3(b)). HEED is chosen since it provides a single-hop SN-to-CH communication, and multi-hop CH-to-CH using a directed diffusion (DD) routing protocol. Moreover, it assumes a nonuniform energy consumption and SN-location knowledge is not necessary for clustering. It takes into consideration the residual energy per node while performing inter- and intracommunication, where nodes with higher residual energy have higher probability to be elected for a data query. And thus, it maintains balanced clusters of these randomly distributed SNs.

In order to record all clusters at the central sink node, we need a data structure associated with each relay's cluster to store coordinates and total number of SNs connected to the RN. We represent this data structure by the set $C(i)$, where i is the vertex index on which the RN is placed. By computing $C(i)$, $\forall\ i \in V$, we can test whether a RN at vertex i is optimal or not by searching for a set that has at least all elements of $C(i)$. In the following, Algorithm 3.2 is running independently at the SNs/RNs by the end of each round to collect residual energy and neighboring SNs per RN. If any change occurs in the resultant gathered information, it will be broadcasted to update the network sink.

Algorithm 3.2: SNs Communication Assignment

1. **Inputs**:
2. A set N of the RNs/SNs' coordinates.
3. **Outputs**:
4. $C(i)$: Set of covered SNs per RN at i.
5. E: remaining energy at SN/RN in cluster i.
6. **Begin:**
7. If SN
8. $E = E_{rem}$ with reference to Equation 3.3
9. Endif
10. If RN
11. $E = E_{rem}$ RN with reference to Equation 3.3

12. Endif
13. C(i):= φ; //list of covered SNs by relay node i
14. foreach RN at vertex i do
15. Compute Pr[i, j] with reference to Equation 3.7
16. If **Pr** ≥ γ_{th}
17. C(i):= $j \cup$ C(i);
18. endif
19. endfor
20. **End**

3.5 Simulation Results and Discussions

Simulation results in this section have been derived from Omnet++ and MATLAB® simulations. Our initial setup in Omnet++ considered the use of a planned 2D-grid deployment of CNs in a randomly deployed sensor network. The assumed virtual grids are squares of side length equal to 200 m. Initially, we considered three techniques for data routing. The first technique, Tech. A, is based on the directed diffusion (DD) routing approach [28], where the data was returned along the same path on which it was arrived. In this technique we assume only sensor nodes constructing the overall network. The second technique, Tech. B, allowed the usage of RNs in order to support the distributed sensor nodes while performing sort of aggregation in order to avoid as much as possible redundancy in transmissions. In the third technique, Tech. C, we use CNs to choose a different data delivery path, compared to the request arrival path. We also let the CNs communicate across the diagonals of the square grid while adapting their transmission power. To increase the probability of reliable data reception at nodes located at different separation distances from the source node, the CNs were allowed to vary their transmit power between 0 dBm to 10 dBm, with specific increment values of {0, 3, 5, 7, 10} dBm as described in Table 3.2. This can show how cognition can improve existing WSNs in terms of the network lifetime. We ran our simulations until first node death for all techniques and found that Tech. B is experiencing almost the same performance as in Tech. A. However, Tech. C is significantly outperforming the other two techniques in terms of network lifetime due to the added CNs in the network, as shown in Figure 3.6. This can be returned to the utilized cognition in

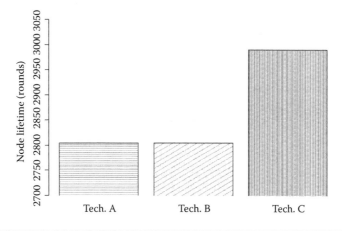

Figure 3.6 Comparison of the three simulated techniques based on total number of transmissions rounds (lifetime).

adapting network resources according to users' requests, where transmission powers are changed according the data-request's source and the current communication link conditions. In fact, the proposed CNs make use of the received feedback about the utilized channel condition and modulation rate to determine the sleep time of each node. This concept has been emphasized more in the following simulation study while assuming realistic parameters' values based on [29] and summarized in Table 3.6 for energy consumption of each node in four modes: Sleep mode, receive mode, active mode (ready to transmit but not transmitting), and transmission mode, as shown in Figure 3.7.

Figure 3.7 displays the mean node lifetime in a WSN using adaptive modulation and adaptive sleep (AMS) versus adaptive modulation (AM) only. In case of AM, the modulation level (parameter M in

Table 3.6 Node Parameters Used in Simulation

PARAMETER	VALUE
Current consumption in Sleep mode: I_{sleep}	1 µA
Current consumption in Receive mode: I_{rx}	20 mA
Current consumption in active mode: I_{ac}	100 mA
Current consumption while transmitting	120 mA
Traffic intensity	90 percent
Log–Normal Shadowing variance (σ)	0, 2 dB, 4 dB, or 6 dB
BER required (QoS)	10^{-4}
RF Bandwidth used	200 kHz

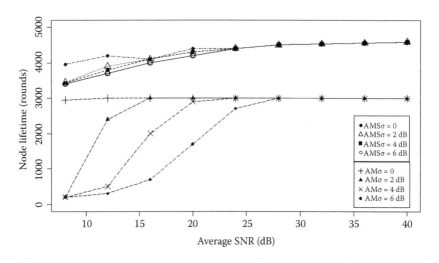

Figure 3.7   Node lifetime using adaptive modulation (AM) vs. adaptive modulation and adaptive sleep (AMS).

M-QAM modulation) is chosen for each packet according to channel condition (i.e., SNR). This case assumes no cognition and the sleep time is predetermined independently from the user requests or any changes to application requirements. Meanwhile, the same figure shows the average node life time using an adaptive modulation scheme combined with a scheduled sleep (AMS) mechanism that adapt based on the CNs feedback. In other words, cognition here is employed at the MAC and PHY layers and sleep times are scheduled according to channel conditions and bit rates. In both cases high traffic patterns (mean packet arrival rate 90 percent) are assumed and simulated using Poisson distributions and log–normal shadowing where shadowing variance takes values from 0 (no shadowing) to 6 dB. The figure shows that the cognitive approach significantly outperforms the noncognitive one in terms of the average node lifetime. This is because M-QAM modulation, when carefully chosen, will require less transmission time, and thus, the cognitive system exploits this information to modify the sleep time accordingly. The improvement made by the cognitive approach is higher for both high traffic intensity and severe channel conditions (i.e., low SNR and high shadowing variance). To mitigate the need for the increment in power transmission in typical WSN deployment strategies in order to maintain a reliable communication channel over prolonged distances, we considered the use of relay nodes (RNs) between the diagonally communicating CNs, which serves as multi-hop communication paths. But adding RNs in the current setup means increasing the total number of nodes deployed in the network, thus increasing the overall hardware cost of the network. So, neither only increasing the transmit power at CNs, nor simply adding extra RNs in the deployment plan, provide elegant solutions (i.e., cognitive deployments) to improving the connectivity, or reducing the energy consumption of the network. From Equation 3.9, we know that the operational cost of CN is higher than that of RNs. This indicates that it is feasible to have more number of RNs than CNs in the network. Hence, we adopt the deployment strategy proposed in Algorithm 3.1 to maintain a lower ratio of CNs to RNs used in the network.

The transmit power of RNs and CNs, and their respective communication ranges are planned based on Table 3.2 and Figure 3.4, and the resulting node deployment plan is as shown in Figure 3.8. This deployment plan was implemented in MATLAB® as described in the following subsections to allow more flexibility with the parameter setting and control over the CN's behavior during the network operation. The simulations were used to study the impact of the node

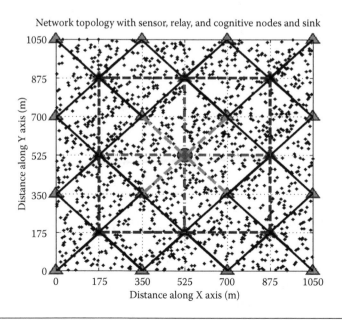

Figure 3.8   ICSN with CNs, RNs, and SN in the network.

deployment and internode interactions on the quality of information (QoI) attributes of latency, reliability, and throughput. Details of the definitions of the QoI attributes and the simulation setup are provided in the following sections.

3.5.1 The Quality of Information (QoI) Attributes

We use latency, reliability, and instantaneous throughput as QoI evaluation metrics for the IEEE 802.15.4 MAC-PHY layer in typical WSNs.

3.5.1.1 Node Reliability (NR) at the Transmitting Node

Node reliability is defined as the probability that a transmitting node is able to successfully deliver a data packet to its next hop neighbor. It is a function of the node's buffer capacity (blocking probability), and the channel conditions (based on the number of nodes trying to simultaneously transmit data) at the time of channel access/data transmission. Thus it inherently reflects upon the link reliability as well. This definition of reliability is based on the work in [30] for low-power nodes in the 802.15.4 PHY-MAC model. We apply the same definition to the cognitive nodes as well, as they will be interacting with sensor and relay nodes in the same setting.

$$NR = ((1 - P_{blocking}) * (1 - P_{c-fail}) * (1 - P_{p-discard})) \qquad (3.12)$$

Where $P_{blocking}$ represents the blocking probability due to a buffer-full condition; P_{c-fail} is the channel access failure probability and $P_{p-discard}$ is the probability that a packet is discarded on reaching the maximum number of retries limit.

3.5.1.2 Instantaneous Throughput (IT) at the Receiving Node

The definition for instantaneous throughput (IT) at a receiving node is based on the work in [29,31] and is applied to both relay and cognitive nodes. It is defined as a ratio of the size of the frame payload at the physical layer (Overhead + application payload) L in bits, over the mean service time M in seconds.

$$IT = L/M \quad \text{(bits/s)} \qquad (3.13)$$

3.5.1.3 Observed Latency (OL) at the Receiving Node The observed latency at a receiving node accounts for delays due to the mean service time at the transmitting node, which is a function of the frame arrival rate. In the following section, we evaluate the node reliability (NR), observed latency (OL), and instantaneous throughput (IT) for the proposed deployment strategy and communication range after initiating our simulation setups in MATLAB®.

3.5.2 Simulation Setup

Using MATLAB® (R2013a), we simulated the deployment plan for a large-scale ICSN with 1500 SNs, 16 RNs, 9 CNs, and a sink over a square target area of side A = 1050 m. The SNs were distributed randomly and uniformly over the target area. CNs and RNs are deployed at fixed, equidistant locations on a two-dimensional square grid as described in the deployment plan, and shown in Figure 3.3(b). The RNs are deployed at the corners of each square grid and the CNs at the center. The sink is deployed at the center of the deployment region. Node connections and interactions are based on a hierarchical ZigBee topology model [32]. The network is built over an IEEE 802.15.4 MAC-PHY simulator based on the work in [30]. Simulation parameters are set based on practical experimental results in [29] as depicted in Table 3.7. The parameters that are varied in our simulation model are listed as follows:

(a) $N_{simultaneous}$: the number of nodes attempting to simultaneously transmit data

(b) Load: Application payload in terms of the size of the MAC frame payload in *bytes*, and

Table 3.7 Simulation Parameters and Values [29,30]

PARAMETER NAME	VALUE
Operational frequency	916 MHz (ISM band)
Data rate	250 kbps
Transmit power	0 dBm for RN; and any value from the set {0, 3, 5, 7, 10} dBm for the CN
Modulation	Phase shift keying (PSK)
Encoding	Non-return to zero
Path loss model	Log–normal shadowing n = 4, σ = 4
SN transmission range	175 m
Application payload size	0–127 bytes

(c) Offered load: per node frame arrival rate expressed as a fraction of the application payload, in *bits/second.*

Impact of varying of these parameters on the QoI attributes of *latency* (OL), *reliability* (NR), and *instantaneous throughput* (IT), and *average throughput* are studied. The maximum and minimum possible values for $N_{simultaneous}$ were identified from the node binding information available at the time of network deployment.

From 10 sets of random deployment of sensor nodes, we found a lower bound of about 10 sensor nodes per CN, and an upper bound of close to 60 sensor nodes per CN, which we used in the simulations. The range of values chosen for the application payload size was 51 to 121 bytes. This range for the MAC frame payload size was chosen based on the size of the data field supported by IEEE 802.15.4 Physical layer packets: 0 to 127 bytes [30]. The range of values for per node offered load was 0 to 1400 bits/second, such that the load could be expressed as a fraction of the application payload, ranging from 0.1 to 1.4 times the size of the application payload.

3.5.3 Simulation Results

From the simulation results shown in Figures 3.9 through 3.12, we analyze the impact of varying $N_{simultaneous}$ on the QoI attributes for different application payload sizes. From Figure 3.9, we see an overall trend of increase in latency as the number $N_{simultaneous}$ increases. Latency increases with increase in application payload size, for any value of

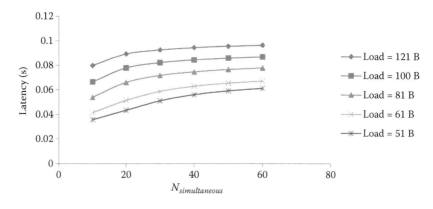

Figure 3.9 Impact of $N_{simultaneous}$ and load on latency.

$N_{simultaneous}$. For instance, we see that the latency for $N_{simultaneous}$ = 10, for a 51 B payload, is less than 0.04 s. Figure 3.10 indicates an overall trend of decrease in reliability with increase in the $N_{simultaneous}$ nodes for any payload size. For a given value of $N_{simultaneous}$, lower reliability values were observed for higher payloads. For instance, when $N_{simultaneous}$ = 20, reliability improves from a value of about 0.4 at a payload of 121 B to about 0.8 at 51 B, which is almost a 50 percent improvement in reliability. At higher values of $N_{simultaneous}$, although the absolute value of reliability is small, the percentage difference between reliability values for different payload sizes remains almost same, especially when compared at 51 B and 121 B payloads.

Figure 3.11 and Figure 3.12 indicate an overall trend of decrease in throughput as the $N_{simultaneous}$ nodes increase. However, both average and instantaneous throughput values are higher for higher application

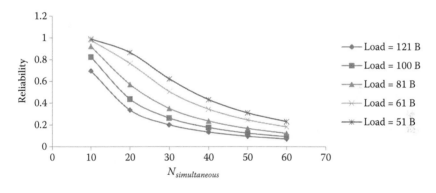

Figure 3.10 Impact of $N_{simultaneous}$ and load on reliability.

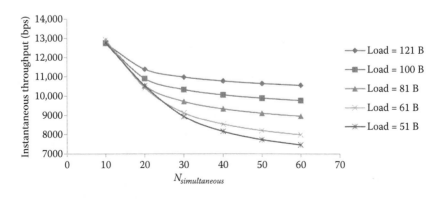

Figure 3.11 Impact of $N_{simultaneous}$ and load on instantaneous throughput.

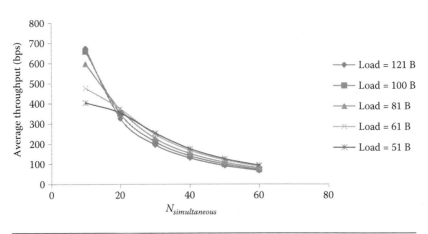

Figure 3.12 Impact of $N_{simultaneous}$ and load on average throughput.

payload sizes especially for values of $N_{simultaneous}$ less than 20 nodes. Thus, controlling the number of nodes that are scheduled for simultaneous transmission to keep it between 10 and 20 nodes helps to improve the network performance in terms of the QoI attributes, even at high application payloads.

Next, we analyze the interdependence of the QoI attributes, and the variation of average wait time at a node as the frame arrival rates increase as shown in Figure 3.13 and Figure 3.14. From Figure 3.13 we can see that IT drops to almost half its value of about 5×10^4 bps for a frame

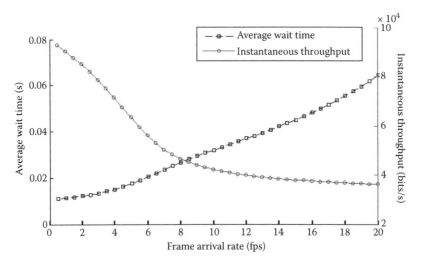

Figure 3.13 Average wait time and instantaneous throughput versus per node frame arrival rate.

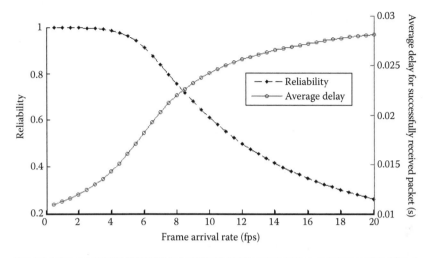

Figure 3.14 Reliability and average delay for successfully received packet versus per node frame arrival rate.

arrival rate of 6 fps, from an original value of 9.8×10^4 bps for a frame arrival rate of less than 1 fps. The average wait time also sees a steeper increase beyond a 5 fps frame arrival rate. So, when increased traffic volumes in the network causes an increase in the frame arrival rate at the network nodes, the performance in terms of the per-node OL and IT degrades. NR follows a similar trend as observed from Figure 3.9. As the frame arrival rate increases beyond 6 fps, NR drops below 0.9 and the average delay for successfully received packets is about 75 percent more; from about 0.01 s at 1 fps arrival rate to 0.018 s at 6 fps. In summary, we observe that the performance of each of the QoI attributes deteriorates with increased frame arrival rates at the nodes. But the link quality also affects the network's performance, apart from the per-node performance. From the observed performance of the QoI attributes in large-scale ICSN applications, we see that there is scope for improving the network performance in terms of the network traffic and resource management. In addition, we were also able to identify how the nodes can be scheduled to achieve better performance in terms of the QoI attributes. Thus, we suggest that cognitive nodes can be introduced to better manage the traffic flows and network resources, in a way that is cognizant of the end user's service requirements. If an application requires high throughput guarantees, then cognitive nodes can exercise their learning mechanism to understand the user requirements and the reasoning mechanism can exercise its control to limit the number of nodes scheduled for simultaneous transmission and to provide the user-desired reliability for the application being serviced. These control instructions from cognitive nodes are passed on to RNs as well and the performance of the entire network can be tuned to meet the application's service requirements.

3.6 Conclusions

WSNs require additional capabilities in terms of being able to understand the network and users' dynamic behavior and store the large volume of generate data in the cognitive ICSN infrastructure. The network is dynamic due to changing application requirements, end user behavior, and network topology. To address these challenges, we proposed the introduction of cognitive nodes which implement learning, reasoning and knowledge representation as elements of cognition, in large-scale ICSN applications. We proposed the cognitive node architecture, and described the interactions among the cognitive elements and how they would impact the network performance. The reasoning mechanism responds to current network conditions that require immediate attention; while the learning mechanism uses the knowledge acquired during the network operation over a period of time, for planning and controlling the response to predicted network conditions. Next, based on the energy consumption and hardware cost of the relay and cognitive nodes, the number and position of RNs and CNs to be used in the network was determined by the deployment plan. The proposed deployment plan was evaluated in terms of its performance of the QoI attributes of latency, reliability and throughput. We found that if the CNs could exercise control over the number of nodes that are scheduled for simultaneous transmission, and keep the number below 20 nodes, the QoI attributes' performance could be maintained at reasonable values for the proposed deployment plan and transmit power capabilities.

Acknowledgment

The author would like to thank his outstanding PhD student, Dr. Gayathri Singh, for her appreciated assistance in achieving this work.

References

1. ITU-T technology watch briefing report series no. 4, "Ubiquitous Sensor Networks," Feb. 2008.
2. I. Stojmenović and S. Olariu, "Data-Centric Protocols for Wireless Sensor Networks," in *Handbook of Sensor Networks* (ed. I. Stojmenovic), pp. 417-456, John Wiley &, Sons, Hoboken, NJ, 2005.
3. F. Al-Turjman and H. Hassanein, "Towards Augmented Connectivity with Delay Constraints in WSN Federation," *Inderscience: International Journal of Ad Hoc and Ubiquitous Computing*, vol. 11, no. 2, pp. 97–108, 2012.
4. A. Al-Fagih, F. Al-Turjman, W. Alsalih, and H. Hassanein, "A Priced Public Sensing Framework for Heterogeneous IoT Architectures," *IEEE Transactions on Emerging Topics in Computing*, vol. 1, no. 1, pp. 133–147, June 2013.
5. Z. Yun, X. Bai, D. Xuan, T. H. Lai, and W. Jia, "Optimal Deployment Patterns for Full Coverage and k-Connectivity (k ≤ 6) Wireless Sensor Networks," *IEEE/ACM Transactions on Networking (TON)*, vol. 18, no. 3, pp. 934–947, 2010.
6. A. Tufail, "Reliable Latency-Aware Routing for Clustered WSNs," *International Journal of Distributed Sensor Networks*, vol. 2012, Article ID. 681273, 6 pp., 2012.
7 X. Cheng, D. Z. Du, L. Wang, and B. Xu, "Relay Sensor Placement in Wireless Sensor Networks," *ACM/Springer Journal of Wireless Networks*, vol. 14, no. 3, pp. 347–355, June 2008.
8. F. Al-Turjman, H. Hassanein, and M. Ibnkahla, "Optimized Relay Placement to Federate Wireless Sensor Networks in Environmental Applications," 7th International Wireless Communications and Mobile Computing Conference (IWCMC), pp. 2040–2045, July 4–8, 2011.
9. F. Wang, D. Wang, and J. Liu, "Traffic-Aware Relay Node Deployment: Maximizing Lifetime for Data Collection Wireless Sensor Networks," *IEEE Transactions on Parallel and Distributed Systems*, vol. 22, no. 8, pp. 1415–1423, 2011.
10. Y. B. Reddy and C. Bullmaster, "Application of Game Theory for Cross-Layer Design in Cognitive Wireless Networks," in *Proc. 6th Int. Conf. Inform. Technology: New Generations, ITNG*, 2009, pp. 510–515.
11. C. Bisdikian, L. M. Kaplan, and M. B. Srivastava, "On the Quality of Information in Sensor Networks," *ACM Trans. Sensor Netw*, vol. 9, no. 4, article 48, July 2013.
12. B. Ahlgren, C. Dannewitz, C. Imbrenda, D. Kutscher, and B. Ohlman, "A Survey of Information-Centric Networking," *IEEE Communications Magazine*, vol. 50, no. 7, pp. 26–36, July 2012.
13. F. Al-Turjman and H. Hassanein, "Enhanced Data Delivery Framework for Dynamic Information-Centric Networks (ICNs)," in *Proc. of the IEEE Local Computer Networks (LCN)*, Sydney, Australia, 2013, pp. 831–838.

14. B. Krishnamachari, D. Estrin, and S. Wicker, "Modelling DataCentric Routing in Wireless Sensor Networks," *IEEE Infocom*, vol. 2, pp. 39–44, June 2002.

15. F. Al-Turjman, H. Hassanein, and M. Ibnkahla, "Towards Prolonged Lifetime for Deployed WSNs in Outdoor Environment Monitoring," *Elsevier Ad Hoc Networks Journal*, vol. 24, no. A, pp. 172–185, Jan. 2015.

16. F. Al-Turjman, "Impact of User's Habits on Smartphones' Sensors: An Overview," HONET-ICT International IEEE Symposium, Kyrenia, Cyprus, pp. 70–74, Oct. 2016.

17. G. Singh, and F. Al-Turjman, "Cognitive Routing for Information-Centric Sensor Networks in Smart Cities" in *Proc. of the International Wireless Communications and Mobile Computing Conference (IWCMC)*, Nicosia, Cyprus, 2014, pp. 1124–1129.

18. F. Al-Turjman, "Information-Centric Sensor Networks for Cognitive IoT: An Overview," *Springer Annals of Telecommunications Journal*, pp. 1–16, 2016. DOI: 10.1007/s12243-016-0533-8.

19. K.-P. Yee, D. Fisher, R. Dhamija, and M. Hearst, "Animated Exploration of Dynamic Graphs with Radial Layout," *Proc. Information Visualization*, 43–50, 2001.

20. Lithium-Based Batteries [online]. Available at http://batteryuniversity .com/learn/article/lithium_based_batteries.

21. X. Fang, S. Misra, G. Xue, and D. Yang, "Smart Grid—The New and Improved Power Grid: A Survey," *IEEE Communications Surveys and Tutorials*, vol. 14, no. 4, pp. 944–980, 2011.

22. F. M. Al-Turjman, H. S. Hassanein, and M. A. Ibnkahla, "Efficient Deployment of Wireless Sensor Networks Targeting Environment Monitoring Applications," *Computer Communications*, vol. 36, no. 2, pp. 135–148, 2013.

23. F. Al-Turjman, H. Hassanein, and M. Ibnkahla, "Quantifying Connectivity in Wireless Sensor Networks with Grid-Based Deployments," *Elsevier: Journal of Network and Computer Applications*, vol. 36, no. 1, pp. 368–377, Jan. 2013.

24. Anonymous, "Pythagorean Theorem," [online]. Available at http://math world.wolfram.com/PythagoreanTheorem.html.

25. R. Saaty, "The Analytical Hierarchy Process—What It Is and How It Is Used," *Elsevier Mathematical Modelling*, vol. 9, no. 3, pp. 161–176, 1987.

26. M. Z. Hasan, F. Al-Turjman, and H. Al-Rizzo, "Evaluation of a Duty-cycled Protocol for TDMA-Based Wireless Sensor Networks," in *Proc. of the International Wireless Communications and Mobile Computing Conference*, Paphos, Cyprus, pp. 964–969, 2016.

27. F. Al-Turjman, "Hybrid Approach for Mobile Couriers Election in Smart-cities," in *Proc. of the IEEE Local Computer Networks (LCN)*, Dubai, UAE, pp. 1–4, 2016.

28. C. Intanagonwiwat, R. Govindan, and D. Estrin, "Directed Diffusion: A Scalable and Robust Communication Paradigm for Sensor Networks," *Proceedings of the 6th Annual International Conference on Mobile Computing and Networking*, ACM, 2000.

29. M.-H. Zayani, V. Gauthier, and D. Zeghlache, "A Joint Model for IEEE 802.15.4 Physical and Medium Access Control Layers," in *Proc. of IEEE Int. Wireless Communications and Mobile Computing Conference* (IWCMC 2011), 2011.

30. M.-H. Zayani, and V. Gauthier, "Usage of IEEE 802.15.4 MAC–PHY Model," available from http://www-public.it-sudparis.eu/~gauthier/Tools /802_15_4_MAC_PHY_Usage.pdf.

31. P. Park, P. Di Marco, P. Soldati, C. Fischione, and K. H. Johansson, "A Generalized Markov Chain Model for Effective Analysis of Slotted IEEE 802.15.4," *IEEE 6th International Conference on Mobile Adhoc and Sensor Systems*, vol. 130, no. 139, pp. 12–15, Oct. 2009.

32. F. Al-Turjman, A. Al-Fagih, H. Hassanein, and M. Ibnkahla, "Deploying Fault-Tolerant Grid-based Wireless Sensor Networks for Environmental Applications," *2010 IEEE 35th Conference on Local Computer Networks (LCN)*, pp. 715–722, Oct. 2010.

4

QUANTIFYING CONNECTIVITY IN WIRELESS SENSOR NETWORKS WITH GRID-BASED DEPLOYMENTS*

FADI AL-TURJMAN,
HOSSAM S. HASSANEIN,
AND MOHAMAD IBNKAHLA

4.1 Introduction

Grid-based deployments [1] have been widely used in Wireless Sensor Networks (WSNs) to achieve significant improvements in terms of the network coverage and connectivity [1,2]. Such deployments can effectively limit the 3D search space of the sensor candidate positions by placing the sensor nodes on well-organized vertices, in regular lattice structures. These vertices can be organized in different structures (e.g., cubes, octahedrons, pyramids) in the 3D space to provide more accurate estimates in terms of the spatial properties of the measured data [3]. In addition, the grid deployment becomes necessary when sensor nodes are expensive and their operation is significantly affected by their positions. Such deployments can have a wide range of applications, such as aircraft health monitoring, pollution and CO_2 flux monitoring, forest fire detection, and gigantic redwood tree monitoring [3,4]. However, these applications may have several deployment challenges in practice such as placement uncertainty and communication irregularity, and may be very critical and require real-time interaction as in the ability to immediately

---

* This article was originally published in *Journal of Network and Computer Applications*. F. Al-Turjman, H.S. Hassanein, M. Ibnkahla, Quantifying connectivity in wireless sensor networks with grid-based deployments, vol. 36, no. 1, pp. 368–377, 2013. Reprinted with permission.

respond to natural disasters [5,6]. Therefore, connectivity between the deployed sensor nodes on the grid and the system sink is of utmost importance to ensure the timeliness delivery of the measured data, and thus, numerous research papers have focused on how to maximize the network connectivity and/or to maintain it as a functional constraint during the network deployment planning [7]. In this research, we quantify the grid-based deployment connectivity and show how resilient it is while assuming practical deployment challenges in the aforementioned applications.

Connectivity in the context of wireless sensor networks (WSNs) is measured by the minimum number of independent paths between every pair of nodes in the deployed network, and this metric is called k-connectivity [8]. With k-connectivity, where k ≥ 2, sensor nodes have a better chance to stay connected with the base station (system sink). Nevertheless, this metric may not be accurate in practice where deployed nodes are subject to positioning errors and other practical challenges.

Practical deployment planning in WSN applications is a challenging problem for three main reasons: (1) Placement uncertainty, (2) communication range irregularity, and (3) 3D setups [9]. Placement uncertainty could occur because of timing errors in the placement, inaccurate distance estimation, unexpected changes in location due to weather conditions such as rain and wind, or wildlife disturbing the deployed nodes. Errors due to human mistakes and low-accuracy deployment strategies, which are used for cost effectiveness, would also amplify the physical placement uncertainty.

Communication irregularity in WSN applications stems from natural or man-made obstacles in the terrain, such as trees, walls, mountains, cliffs, and so on, in addition to extreme weather conditions in outdoor applications, which may affect the propagated signals [10]. Different power levels of the received signal may be observed at different positions, even when these positions are equally distanced from the transmitter [11]. Assuming a regular shape for the communication range is hence not plausible in practice. Flat and obstacle-free terrains are also highly unlikely in outdoor applications. Consequently, significant efforts have focused on modeling irregular radio propagation and noisy communication channels in the past few years [10]. However, only a few of these efforts have been applied in the presence of placement uncertainty.

Moreover, in certain applications, the vertical position of the sensor node is as important as the longitude and the latitude of its placement.

In these applications, a 3D network deployment is required. For instance, in monitoring the gigantic redwood trees in California, some experiments required sensor placements at different heights on these trees spanning a range of several tens of meters [12]. There is also increased interest in 3D environmental applications such as CO_2 flux monitoring and imagery, where sensors are placed at different vertical levels to fulfill the monitoring coverage and accuracy requirements. It is worth mentioning that there are some attempts towards the 3D deployment. As an example, authors in reference [13] studied the effects of sensing and communication range on connectivity in 3D space. While grid-based deployment in 3D space has been used in various applications and several attempts targeted the 3D connectivity, the average connectivity of the grid when probabilistic communication models and placement uncertainties are assumed has not been investigated yet. Considering properties of the grid connectivity during the deployment planning will result in more efficient design decisions in terms of the required grid-edge length and the maximum range of errors the grid can tolerate to satisfy specific connectivity requirements. Consequently, unnecessary cost (extra useless nodes) and traffic overhead (unnecessary retransmissions) can also be reduced in the designed sensor network which is assumed to be cost and energy constrained.

To this end, in this chapter, we examine connectivity properties of the 3D grid-based deployment under communication irregularity and placement uncertainty in accordance with reference [9]. We assume deployed nodes are subject to random errors in physical positioning. In addition, we assume a probabilistic communication range, which has an arbitrary irregular shape in 3D space (due to the presence of obstacles, and signal attenuation and reflection).

The major contributions of this chapter are as follows: (1) We provide a solid analytical derivation for the node positioning errors and the communication range irregularity on the grid; (2) we explore a more practical connectivity setting for the deployed sensor nodes by considering their placement in the 3D space; (3) assuming such practical aspects of the deployment, we propose a generic approach to evaluate the average connectivity in 3D grid-based deployments and this approach is applicable to various grid shapes and different kinds of random error distributions; and (4) based on the proposed approach, other design problems, such as the maximum allowed placement error and maximum grid-edge length to satisfy a connectivity requirement, are also addressed.

The remainder of this chapter is organized as follows. Section 4.2 outlines the related work. Section 4.3 describes the practical communication and error models assumed in this chapter. A generic approach to measure node connectivity in grid-based deployments is proposed in Section 4.4. In Section 4.5, we verify the correctness of the proposed approach through simulation results, in addition to analyzing practical design issues of the grid-based deployment through numerical results. Finally, we conclude this chapter in Section 4.6.

4.2 Related Work

In grid-based WSN deployments, practical errors can affect the coverage or connectivity of the grid. The coverage of the grid has been extensively studied in the literature and several approaches have been explored to overcome its experienced problems. In addition, structural properties of the grid have been investigated in order to solve coverage problems before their occurrence. In reference [1], triangular grid-based deployment for coverage, when sensor placements are perturbed by bounded random errors around their corresponding grid vertices, is studied. The random errors are modeled by uniform displacements inside error disks of a given finite radius on a 2D plane. The average coverage of the sensing field is derived as a function of the length of the grid edges and the radius of the random error disks. It was shown how resilient the grid coverage is to different random error distributions in the sensor displacement.

Extensive work has also been applied to overcome connectivity problems and failures in grid-based deployments. Nonetheless, the majority of this work is proposed to repair connectivity problems after their occurrence either by using node redundancy [8,14] or node mobility [15–17]. Node redundancy [8] is used to overcome disconnected networks. Redundant nodes, which are not being used for communication or sensing, are turned off. When the network becomes disconnected, one or more of the redundant nodes is turned on to resume the network connectivity. In reference [18], a distributed recovery algorithm was developed to address k-connectivity requirements, where k is equal to 1 and 2. The idea was to identify the least set of nodes that should be repositioned in order to reestablish a particular level of connectivity. One shortcoming of this technique is the requirement for extra nodes, which may not be cost effective in certain applications. Also, when some redundant nodes fail, it may no longer be possible to repair the network connectivity. Meanwhile, node mobility can also be used to maintain the network connectivity [15]. Typically, mobile nodes are relocated after deployment to carry data between disconnected partitions of the network. Providing mobility-aware connectivity management decisions with limited overhead, has also been considered recently in [19]. However, relocating nodes without considering grid-connectivity properties and characteristics can have severe effects on the direction of movement and the choice of the most appropriate node to be moved.

Although the above techniques aid in repairing connectivity problems, they do not address the sources of these problems. Our work presents a different approach toward addressing connectivity problems, which also complements the work of the aforementioned techniques. This approach aims at addressing key sources behind connectivity problems and predicts a better deployment planning in grid-based WSNs. Unlike the aforementioned techniques, we propose a measuring metric for connectivity properties in practice and under realistic scenarios such as inaccurate node positioning and communication irregularity. Thus, more efficient connectivity planning and maintenance can be achieved without the use of redundant nodes or expensive mobility-dependent techniques. Consequently, a careful consideration of the proposed metric would eliminate the majority of undesired connectivity problems and costly repairs while the wireless sensor network is in process.

4.3 System Models

In this section, specific models for the assumed communication irregularity and placement errors along with the utilized network model are detailed.

4.3.1 Network Model

In this chapter, a flat network architecture in which all of the deployed (sensor) nodes are of the same type and have the same communication range is assumed. The sensor nodes have a fixed and limited transmission range, and communicate periodically with the base station in a multi-hop fashion. Topology of these sensor nodes is modeled as a graph $G = (V, E)$, where V is the set of n_c candidate grid vertices, and E is the set of edges in the graph G. We remark that the deployment of sensor nodes in this research is independent of the underlying medium access control (MAC) protocol. We assume a transmission rate limit T_r for each node during a single time unit (measured in hours). This limit can be adjusted to comply with any MAC protocol. Figure 4.1 depicts the 3D grid model assumed in this chapter, where the grid edge length is supposed to be proportional to the theoretical sensor nodes transmission range r. We remark that our deployment planning is applicable for other types of grid models, not only the cubical one. On the cubic grid model, each sensor node (SN) is placed on the closest vertex to the phenomenon of interest for more accurate estimates in terms of the spatial properties of the collected data. The base station placement is based on the application requirements in a fixed position and it is the data sink for the system. Next we seek a mathematical formula that quantifies the connectivity level between the deployed sensor nodes and the system sink (base station [BS]). However, this work is not limited to single base-station networks as we will see in Section 4.4.

4.3.2 Communication Model

In practice, the signal level at distance r from a transmitter varies depending on the surrounding environment. These variations cause an irregular communication range (Figure 4.1), and are captured

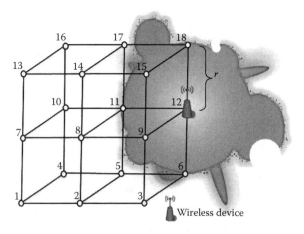

Figure 4.1   An arbitrary shape of the communication range in 3D space.

through the so called log–normal shadowing model. According to this model, the signal level at distance r from a transmitter follows a log–normal distribution centered on the average power value at that point [20]. Mathematically, this can be written as

$$P_d(d) = P_s - P_{loss}(d) = P_s - P_{loss}(r_0) - 10n \log\left(\frac{r}{r_0}\right) + \chi, \qquad (4.1)$$

where P_s is the transmission power, $P_{loss}(r_0)$ is the path loss measured at reference distance r_0 from the transmitter, n is an environment dependent path loss exponent, and χ is a normally distributed random variable with zero mean and variance σ^2, i.e., $\chi \sim N(0, \sigma^2)$. With the aid of this model, the probability of successful communication between two nodes separated with a distance r can be calculated as follows. Assume P_{min} is the minimum acceptable signal level for successful communication between a source S and a destination D separated by distance r. The probability of successful communication is $\rho[S, D] = \text{Prob}[Pd(r) \geq P_{min}]$. After some mathematical manipulations, $\rho[S, D]$ can be written as

$$\rho[S, D] = Q\left(\frac{P_{min} - P_s - P_{loss}(r_0) - 10n \log\left(\frac{r}{r_0}\right)}{\sigma}\right), \qquad (4.2)$$

where $Q(\cdot)$ is the Q-function defined as

$$Q(x) = \frac{1}{\sqrt{2\pi}} \int_x^\infty e^{-t^2/2} \, dt. \qquad (4.3)$$

Thus, the probabilistic connectivity ρ is not only a function of the distance separating the sensor nodes but also a function of the surrounding obstacles and terrain, which can cause shadowing and multipath effects (χ). The ability to communicate between two wireless devices is defined as follows:

Definition 4.1 (Probabilistic Connectivity): Two nodes (devices) separated by a distance d are probabilistically connected with a threshold parameter τ $(0 \leq \tau \leq 1)$, if $\rho \geq \tau$. ∎

4.3.3 Placement Uncertainty Models

Based on the utilized deployment technique, device placement uncertainty can be either bounded or unbounded. In bounded uncertainty, positioning errors are expected to be within a predetermined range. In contrast, the range is unpredictable in unbounded placement errors. Bounded errors may have various random distributions including bounded Normal, bounded uniform, and so on. Similarly, unbounded errors may have various distributions such as the uniform and Normal distribution. For simplicity, and without loss of generality, bounded *uniform* and unbounded *Normal* errors are assumed in this chapter.

4.3.3.1 Bounded Uniform Errors   Assume the deployed device is placed incorrectly at any point inside a bounded error sphere E_g of radius R surrounding the targeted grid vertex. Since the communication range in reality has the arbitrary shape depicted in Figure 4.1, the following lemma holds.

Lemma 4.1: Every surface in a 3D space is a set of infinite spherical subsurfaces.

   Proof. With reference to [21], assume S is any surface in the 3D space. M is a point that belongs to S. Since all points which belong to a spherical surface have an equal distance (radius) r to a single point called the sphere center, the spherical surface becomes a point when r goes to zero. Hence, every point in the 3D space represents a sphere of radius $r\,(= 0)$. Consequently, every point M belongs to the surface S is a sphere of radius $r\,(= 0)$. The total number of spheres constructing the surface S decreases as more sets of adjacent (nonseparated) points, which belong to a single sphere surface, are found.   ■

   Now, using Lemma 4.1, we can assume that the arbitrary shape in Figure 4.1 is constructed by a set of spheres as depicted in Figure 4.2. These spheres vary in number and size from one arbitrary shape to another. To calculate the probability of communication between the deployed device and its neighbors, assume e denotes the event that a device at vertex u of the grid is connected to the device placed in the error sphere E_g. Assume the completed sphere surface of the intersecting subsurface with sphere E_g is O_u in Figure 4.3, where O_u is the largest sphere surrounding the node placed at vertex 12 in Figures 4.1 and 4.2.

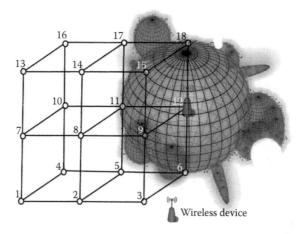

Figure 4.2 Spheres constructing the curved surface of the arbitrary shape in 3D space.

We point out that the adopted log–normal shadowing model, in Section 4.3.2, gives the irregular communication range shape experienced in practice. However, just up to a relatively close distance from the transmitter ($\leq r$) we can assume that connectivity is guaranteed [22]. This guaranteed communication range is indicated by the sphere centered at vertex 12 in Figure 4.2 and called O_u in Figure 4.3.

Then, the probability of e is equal to probability of vertex u being separated from the deployed device in E_g by a distance less than or equal to r. This is satisfied if the deployed device is placed within the

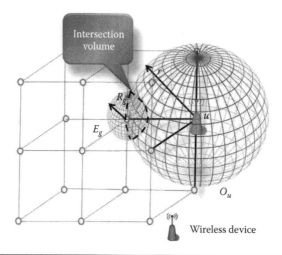

Figure 4.3 Deployed devices are placed within an error sphere E_g which intersects the communication sphere of the device placed at u.

intersection of the sphere O_u centered at u, and the error sphere E_g centered at the grid vertex g. As a result,

$$P(e) = \frac{volume(E_g \cap O_u)}{\frac{4}{3}\pi R^3} \qquad (4.4)$$

Assuming that x axis is passing through the centers of O_u and E_g, and y axis is passing through the center of O_u, we introduce the following lemma.

Lemma 4.2: Intersection volume between the error sphere and the sphere of the intersecting arbitrary subsurface is given by

$$volume(E_x \cap O_u) = \pi \left[\int_{r-R}^{r\cos\theta} [R^2 - (x-r)^2]dx \right.$$
$$\left. + \int_{r\cos\theta}^{r} (r^2 - x^2)dx \right] \qquad (4.5)$$

Proof: Assume a schematic drawing of O_u and E_g in a 2D plane as depicted in Figure 4.4. The intersection volume in Figure 4.3 is then

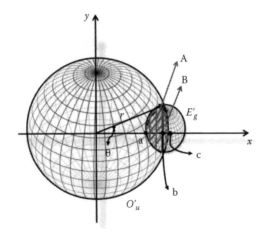

Figure 4.4   Cross-sectional circles of the spheres O_u and E_x in a 2D plane.

generated by rotating the two shaded areas A and B, in Figure 4.4, 180° around the x axis. Hence,

$$volume(E_x \cap O_u) = volume_A + volume_B \qquad (4.6)$$

In order to calculate the volumes generated by A and B, we integrate the corresponding equations of circles O'_u and E'_g from a to b and from b to c, respectively. Since the equation of a circle of radius i centered at point (h, k) is

$$i^2 = (x - h)^2 + (y - k)^2 \qquad (4.7)$$

Assuming O'_u is centered at the point of origin and E'_g is centered at $(x, 0)$, we get

$$volume(E_x \cap O_u) = \pi \int_a^b [R^2 - (x-r)^2] dx + \pi \int_b^c (r^2 - x^2) dx \qquad (4.8)$$

where $a = r - R$, $b = r \cos \theta$, and $c = r$. Therefore, the intersection volume in Figure 4.3 is equivalent to the volume in Equation 4.5. ∎

As the volume of the intersecting spheres has a closed form proposed in Lemma 4.2, we can efficiently compute $P(e)$ for any grid vertex g and its neighboring vertices in the presence of bounded errors. The following theorem is a direct result from Lemma 4.2.

Theorem 4.1: If device placement is subject to uniform bounded random errors, then*

$$P(e) = \frac{\pi \left[\int_{r-R}^{r \cos \theta} [R^2 - (x-r)^2] dx + \int_{r \cos \theta}^r (r^2 - x^2) dx \right]}{\frac{4}{3} \pi R^3} \qquad (4.9)$$

---

*$\theta = \cos^{-1}\left(1 - \dfrac{R^2}{2r^2}\right)$

4.3.3.2 Unbounded Normal Errors In this section, we assume independent trivariate Normal errors in a 3D space around the targeted grid vertex (as the error is unbounded, there is no specific radius for the error sphere). We assume that the error distribution has a mean of zero and is spherically symmetric in three-axis Cartesian coordinates, centered at the grid vertex. Hence the probability density function (PDF) is

$$f(x, y, z) = \frac{1}{(2\pi\sigma^2)^{3/2}} e^{-(x^2+y^2+z^2)/2\sigma^2} \tag{4.10}$$

where σ^2 is the variance. In order to calculate the probability of connectivity between a sensor node placed at the point (u,v,q) and another sensor node placed inaccurately at $(x,y,z)^*$ and separated by a distance d_i, assume e denotes the event that these two sensors are connected. Then, the probability of connectivity between these two sensors, when they are subject to random Normal errors, is given by

$$P(e) = \int\int\int\int P_c(d_i, \chi) f(x, y, z) d\chi\, dx\, dy\, dz \tag{4.11}$$

By plugging Equation 4.2 and 4.10 into Equation 4.11, we obtain the following form of the probability of event e:

$$P(e) = \int\int\int\int \frac{K}{(2\pi\sigma^2)^{3/2}} e^{-\left((x-x_i)^2+(y-y_i)^2+(z-z_i)^2\right)/2\sigma^2}$$
$$e^{-\chi d_i^{\gamma/2}} d\chi\, dx\, dy\, dz \tag{4.12}$$

where

$$\int\int\int_{R^3} \frac{1}{(2\pi\sigma^2)^{3/2}} e^{-(x^2+y^2+z^2)/2\sigma^2} dx\, dy\, dz = 1 \tag{4.13}$$

---

* The point (x,y,z) is located anywhere beside the targeted grid vertex (x_i, y_i, z_i).

The form in Equation 4.12 can be further simplified by completing the square and solving the integrand for a specific value of the path loss γ (= 2), obtaining

$$P(e) = \int \frac{K}{1+2\chi\sigma^2} e^{\frac{-\chi\left((u-x_i)^2 + (v-y_i)^2 + (q-z_i)^2\right)}{1+2\chi\sigma^2}} d\chi \quad (4.14)$$

Based on Equation 4.14, $P(e)$ decreases dramatically as the distance between the grid vertices increases and the surrounding environment becomes more crowded by obstacles. Accordingly, we achieve the following theorem.

Theorem 4.2: If the device placement is subject to unbounded random Normal errors, then the probability of two sensor nodes, deployed on adjacent vertices on the grid, being connected to each other is calculated by Equation 4.14. ∎

4.4 Quantifying the Grid Connectivity

Grid-based deployment in 3D space has been used in various WSN applications. Nevertheless, more efficiency in quantifying the grid connectivity is required, especially when communication irregularity and placement uncertainty are assumed.

4.4.1 Generic Approach

In this section, we propose a *generic approach* to derive the *average connectivity percentage* of the deployed sensor nodes on the 3D grid with various types of placement errors including random bounded and unbounded errors. We first state two definitions that we use in the derivation.

Definition 4.2 (Neighboring Set): The set of sensor nodes placed on grid vertices and connected to a common vertex at the center via a single grid edge. Each node of this set is not included in any other neighboring set. ∎

Note that the number of vertices (sensor nodes) in a neighboring set depends on the grid shape. For example, in a cubic grid a single neighboring set will have six vertices as shown in Figure 4.5.

Definition 4.3 (Neighboring Set Connectivity): The percentage of sensor nodes in a single neighboring set, which can communicate with the sink node via single or multiple hops. ∎

In the following, we derive the connectivity of a single neighboring set (V) mathematically. Suppose we have N sensor nodes distributed on the 3D grid vertices and each sensor node has a probabilistic communication range as shown in Figure 4.1. Let C_j denote the percentage of V, centered at vertex j, being connected to at least one node x_i. Assume x_i is a sink node, or a sensor node connected to the sink node via single or multiple hops. Then, the *neighboring set connectivity* is the expectation of C_j and is calculated by

$$E[C_j] = \sum_V \frac{1}{v} p(u) \qquad (4.15)$$

where $p(u)$ is the probability that the node at vertex $u \in V$ is at least connected to x_i and v is the number of neighboring grid vertices in the set V (e.g., $v = 6$ in cubic grid). Let e_i (where i is the vertex index

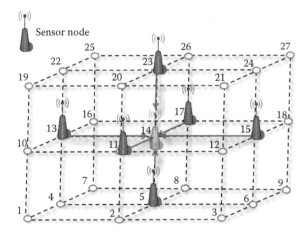

Figure 4.5 An example of a single neighboring set in cubic grid. The neighboring set of the centered node at vertex 14 consists of 6 sensor nodes distributed at vertices 5, 11, 13, 15, 17, and 23.

on the 3D grid and may have a value from *1* to *N*, inclusive) denotes the event that the node at vertex *u* is connected to x_i under random possible errors (bounded/unbounded) in placement. Then we have,

$p(u) = P\left(\bigcup_{i=1}^{N} e_i\right)$. Using the addition rule [20], we have

$$p(u) = \sum_{i=1}^{N} P(e_i) - \sum_{i,j:i<j} P(e_i \cap e_j)$$

$$+ \sum_{i,j,k:i<j<k} P(e_i \cap e_j \cap e_k) + \ldots + (-1)^{N+1} P\left(\bigcap_{i=1}^{N} e_i\right)$$

(4.16)

and using the multiplication rule [20], we get

$$p(u) = \sum_{i=1}^{N} P(e_i) - \sum_{i<j} P(e_i|e_j)P(e_j)$$

$$+ \sum_{i,j,k:i<j<k} P(e_i)P(e_j|e_i)P(e_k|e_i \cap e_j) + \cdots + (-1)^{N+1}$$

$$P(e_1)P(e_2|e_1)P(e_3|e_1 \cap e_2)P(e_4|e_1 \cap e_2 \cap e_3)..$$

$$P(e_N|e_1 \cap e_2 \cap .. \cap e_{N-1})$$

(4.17)

where $P(e_i)$ is calculated from Equations 4.9 and 4.14 in Theorems 4.1 and 4.2 for random bounded and unbounded errors, respectively. And $P(e_i|e_j)$, which is the probability of the node placed at vertex *u* being connected to x_i given that it is connected to x_j under random possible errors, is determined based on the utilized deployment technology. For example, if we are using an airplane to drop the sensor nodes on their specified locations, then based on the speed and wind directions we can estimate the value $P(e_i|e_j)$. However, this is not the usual case in deterministic grid-based deployment, where e_i is often independent of e_j (i.e., $P(e_i|e_j) = P(e_i)$), as will be presented subsequently. We remark that in typical grid deployments, even with random errors, a given sensor node at vertex *u* will be connected with a non-negligible probability only by those sensors that are the closest to *u* (i.e., neighboring set of *u*). Conversely, the probability of being connected to sensors not in the neighboring set of u may be taken to be negligible. Therefore, many terms in Equation 4.17 can be safely set to zero. By computing the summation of Equation 4.17, for all $u \in V$ over v as noted by Equation 4.15, we achieve the single *neighboring set connectivity*. To evaluate the grid connectivity, we compute the *average connectivity percentage* (ACP) of the deployed network.

Definition 4.4 (Average Connectivity Percentage [ACP]): The arithmetic average of all neighboring sets on the grid is the average connectivity percentage of the deployed WSN.   ∎

Assuming N_s is the total number of neighboring sets of the deployed grid, the arithmetic average is

$$Arithmetic\ Average = \frac{\sum_{j=1}^{N_s} E[C_j]}{N_s} \tag{4.18}$$

We remark that this ACP definition is applicable even in the presence of multiple sink nodes, where the ACP is computed by averaging the individual sink's ACPs.

Theorem 4.3: All of the deployed sensor nodes are connected to the sink node if and only if the ACP is equal to one.

Proof. To prove by contradiction, assume that the ACP is equal to one and not all sensors are connected to the sink node. Then, there is at least one sensor which is not connected to the sink. Based on Definition 4.3, there will be at least one neighboring set with average connectivity less than one. Hence, using Definition 4.4, the ACP of the grid is not equal to one which contradicts the assumption.   ∎

In order to illustrate the proposed generic approach further, consider the following example.

Example 4.1: Assume a single system sink and 34 sensor nodes (i.e., $N = 34$) placed on grid vertices as shown in Figure 4.6. Let all sensor nodes be placed exactly on the grid vertices and, for simplicity,

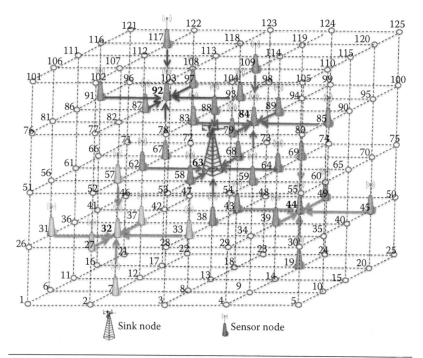

Figure 4.6   Schematic drawing for the grid-based deployment in Example 4.1.

apply the fixed binary communication disk used in reference [7] such that the grid edge length is less than or equal to r_d. In this case, we have five neighboring sets (i.e., $N_s = 5$) centered at vertices 32, 44, 63, 92, and 84. In each neighboring set, v is equal to 6. In order to calculate the *ACP* of the deployed nodes in Figure 4.6, the expected connectivity percentage of each neighboring set is examined. Starting by the sink neighboring set centered at vertex 63, we can see how all of the neighboring nodes of the sink (positioned at vertices 58, 64, 68, 62, 38, and 88) are connected to the sink directly because they are within its range. Hence, $p(u)$ of the sink neighboring nodes is equal to 1 and the expected connectivity percentage $E[C_{63}]$ of the neighboring set centered at the sink is equal to 100 percent. Similarly, by examining the connectivity of the neighboring sets centered at vertices 32, 44, 92, and 84, we find that all of their nodes are connected via single or multiple hops to the sink node. As a result, their expected connectivity percentages will always be equal to 100 percent and thus, the ACP calculated in Equation 4.18 will be equal to 1, as stated in Theorem 4.3. ∎

Based on Example 4.1, we can see how resilient the grid-based deployment is to connectivity failures. For instance, the percentage $E[C_{63}]$ will stay 100 percent even if the communication channels between the sink node and its neighbors placed at vertices 38, 58, 68, and 88 in Figure 4.6 are not available. This is due to the ability of the nodes placed at these vertices to reach the sink node via the sensor nodes placed at vertices 62 and 64. In addition, assuming all possible paths between one of the sensor nodes placed on the grid and the sink node are disconnected, the overall ACP connectivity of the deployed grid-based network will stay very high based on Equation 4.18. It is worth noting that the connectivity degree of the cubic grid deployment proposed in Example 4.1 is equal to six for all sensor nodes except those at the grid boundary where the connectivity degree becomes equal to one. Therefore, dense deployment is required at the grid boundary for a better connectivity degree. Although this example considers ideal grid deployment, it describes the core idea of our generic approach and provides an upper bound on the grid connectivity.

4.4.2 Theoretical Analysis

In Section 4.4.1, we proposed the generic formula that evaluates the ACP of the grid. This formula is applicable to several deployment scenarios, in which the deployment errors occur simultaneously and they may be dependent or independent based on the accuracy of the utilized instruments and technologies in the deployment process. Accordingly, in this section, we examine the ACP of the grid when the events e_i $(i = 1,.., N)$ are mutually exclusive and the deployment errors are assumed independent. Assuming that the events are mutually exclusive can further simplify Equation 4.16 as follows

$$p(u) = \sum_{i=1}^{N} P(e_i) \qquad (4.19)$$

Nevertheless, this assumption is not accurate in practice as the deployed sensor node is most often connected to several other nodes at the same time. Hence, the events e_i are not mutually exclusive except for a very specific case where every deployed sensor node is connected to only one node. On the other hand, deployment errors occur, usually, independently in grid-based deployments due to the placement of each sensor node on the associated grid vertex in an individual manner. Assuming that the deployment errors (bounded/unbounded) are independent means that the e events are mutually independent. And thus Equation 4.17 is reformulated as follows

$$p(u) = \sum_{i=1}^{N} P(e_i) - \sum_{i,j:i<j} P(e_i)P(e_j)$$

$$+ \sum_{i,j,k:i<j<k} P(e_i)P(e_j)P(e_k) + \cdots + (-1)^{N+1}\left(\prod_{i=1}^{N} P(e_i)\right)$$

$$(4.20)$$

For the following discussion, we will assume that the deployment errors are independent, and will therefore use Equation 4.20 to be plugged in Equation 4.15, and thereafter in Equation 4.18.

4.5 Discussion and Numerical Results

In this section, we assess the performance of grid-based deployments in terms of average connectivity (ACP) in the presence of bounded uniform and unbounded placement errors. In both cases we validate our approach using MATLAB® simulation results. Then we discuss properties of the grid connectivity based on some numerical results which have been validated through extensive simulations as well.

We simulate 1000 deployment instances which are randomly generated on the vertices of a cubic grid, and hence, $v = 6$. The grid vertices are distributed in a 900*900*200 (m³) space. As the distance separating the adjacent vertices is equal to the theoretical transmitting range r (= 100 (m)), each simulated deployment can have up to 109 sensor nodes (i.e., $N = 109$) with only one base station (i.e., flat topologies are assumed). Random bounded and unbounded errors are applied on these sensor nodes while they are placed on grid vertices. A linear congruential random number generator is used to distribute up to 109 sensor nodes in the 1000 deployment instances, in addition to generating random errors in the placement coordinates. Based on experimental measurements [23], we set the communication model variables and simulator parameters to be as follows: $\gamma = 2$, $P_{min} = -104$ (dB), $\tau = 60\%$, $T_r = 64$ (byte/hr), and $n_c = 110$ (vertex). And the χ random variable in Equation 4.2 is assumed to be a mean of 0 and variance of 10. Each simulated experiment is repeated 1000 times and the average results hold a confidence interval no more than 2 percent of the average at 95 percent confidence level. Three main parameters are used in this simulation: (1) The error sphere radius R, (2) the variance of the Normal errors σ^2, and (3) the grid edge length L. We chose these parameters as they are key factors in reflecting the range of bounded and unbounded placement errors, and have the major effect on the required average connectivity (ACP) of the grid-based deployment.

We use three main performance metrics: (1) The average connectivity percentage (ACP), (2) the maximum grid edge length, and (3) the maximum displacement error. The ACP reflects the grid-connectivity property under varying placement errors and indicates the efficiency of the deployment plan, as well. The maximum grid edge length is a measurement of the farthest distance separating sensor nodes placed on two adjacent vertices. It is the most influential factor in the design of the grid shape and dimensions. The maximum displacement error is used to reflect the reliability of the deployed grid-based network under practical placement uncertainty. It indicates the maximum allowable error in the placement to satisfy specific connectivity requirements.

4.5.1 Grid Connectivity with Bounded Uniform Errors

In this scenario, the ACP of the deployed sensor nodes is evaluated as a function of the error sphere radius R and the length of the grid edge L. We first study how the radius of the error sphere impacts the average connectivity. Figure 4.7 depicts the impact of R on the ACP with L values varying from 80 to 100 (m) and R values varying from 10 to 40 (m). The results from the theoretical derivations in Section 4.4 and from the simulations match very well, and hence the correctness of our general approach is validated. In addition, we observe how the increase in the radius of the error sphere degrades the grid's ACP. As the grid edge length increases, the grid ACP decreases as well. This is because of the distance consideration in the probabilistic

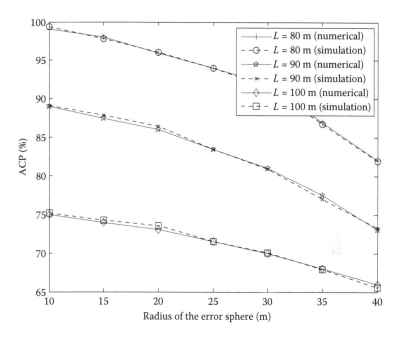

Figure 4.7   Connectivity percentage vs. radius of the error sphere for varying grid-edge lengths.

connectivity model considered in this article. Furthermore, we have observed that having a non-zero error sphere radius may not significantly affect the ACP as long as the grid edge length is short enough. This is because the highest ACP values are achieved with the shortest tested edge length ($L = 80$). Figure 4.7 shows a very appealing feature of the grid-deployment: The network ACP is rather resilient to random bounded errors. As an example, when $L = 80$ and $R \le 40$ (m), the ACP is no less than 80 percent of the ideal connectivity generated with $R = 0$. In addition, the ACP at different L values monotonically decreases with R for the considered probabilistic communication ρ. This is a useful characteristic to solve for the maximum R allowed for a given L and a given connectivity requirement. For example, if the required ACP is 80 percent of the ideal connectivity calculated by the error-free grid deployment, and L is 90 (m), the maximum R should not be greater than 30 (m). Note that when $L = 100$ (m), the ACP cannot reach 95 percent no matter how small R is.

We next investigate how the grid edge length L impacts the ACP in cases where R is set to 10, 20, and 30 (m). We vary the value of L from 75 to 100 (m) in each experiment. The ACP of the deployed nodes as a function of L is then depicted in Figure 4.8. From that figure, we can conclude that as the placement uncertainty is increasing (from 10 m to 30 m) the average connectivity is decreasing. This assures the undesirable effects of placement errors on connectivity as we claimed earlier in this research. Moreover, and as shown in Figure 4.8, the average connectivity is a monotonically decreasing function of L. This assures the effects of the communication irregularity on the average connectivity. That is because as the distance between the transmitter and the receiver (L) is increasing the connectivity is degraded, although this distance (L) is still within the theoretical communication range r (= 100 m) of the communicating devices. Also, this characteristic can be used to solve another practical problem concerning finding the maximum L for a given R, in order to satisfy a specific ACP requirement.

In reference to Figure 4.8, we solve this problem when the required ACP is 85 percent, 80 percent, and 75 percent, and the results are plotted in Figure 4.9. Note that while R increases, the maximum L for the multiple ACP levels decreases. This decrease is expected in order to maintain the required ACP level while the error sphere radius R is

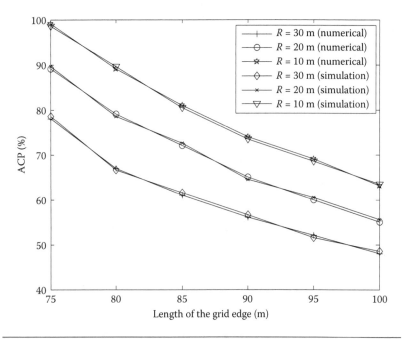

Figure 4.8   Average connectivity percentage vs. length of grid edge for varying error sphere radius.

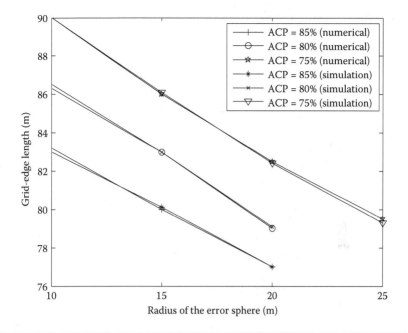

Figure 4.9 Maximum grid's edge length vs. radius of the error sphere to satisfy different connectivity levels.

increasing. Again, results from theoretical derivations in Section 4.4 and from simulations match very well in Figure 4.8 and Figure 4.9, and hence the correctness of the proposed approach is validated.

4.5.2 Grid Connectivity with Unbounded Normal Errors

In this scenario, the ACP of the grid is determined by the variance σ^2 of the Normal distribution and the length of the grid edge L. Figure 4.10 depicts the impact of σ^2 on the ACP with L values varying from 80 to 100 (m) and σ^2 values varying from 0 to 6. Again, numerical results obtained from the generic approach in Section 4.4 match very well with the simulation results, and hence the correctness of our approach is again validated in case of unbounded placement errors.

In accordance with Figure 4.7, Figure 4.10 shows a positive feature of the grid-deployment that the network ACP is also rather resilient to random Normal errors. For example, when $L = 80$ (m), the ACP only drops from 96 percent to 94 percent when σ^2 increases from 0 to 2. The ACP of the grid-based deployment is also a monotonically

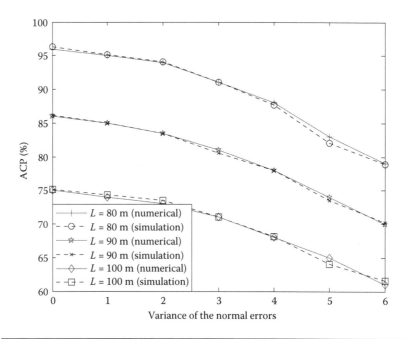

Figure 4.10 Average connectivity percentage vs. variance of the normal distribution for varying grid-edge lengths.

decreasing function of σ^2, and this is a useful characteristic to find the maximum σ^2 allowable for a given L and a given connectivity requirement. In reference to Figure 4.10, Figure 4.11 plots the maximum σ^2 as a function of L. We observe that for ACP to equal 90 percent the maximum variance should not be more than 3.5 and the grid edge length should not exceed 80 (m). In case of ACP to equal 80 percent, the maximum variance can reach 4.7 and 2 for L = 80 and L = 90 (m), respectively. Hence, the highest error variance is a decreasing function of L. This is expected in order to maintain the required ACP level while the grid-edge length L is increasing.

We also study how the distance L impacts the ACP in cases where σ^2 is set to 0, 2, 4, and 6, by varying the value of L from 80 to 100 (m). The ACP of the grid as a function of L is presented in Figure 4.12. Unsurprisingly, the ACP of the grid decreases when the grid edge length increases although this distance (L) is still within the theoretical communication range r (= 100 m) of the communicating devices. This assures the effects of the communication irregularity on the average connectivity even with a nonuniform placement error. As the

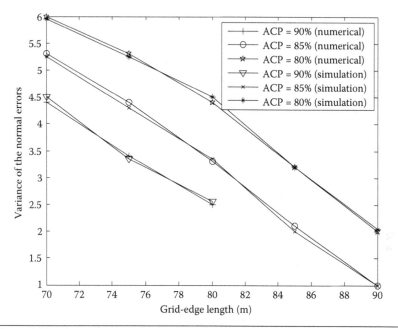

Figure 4.11 Maximum variance of normal distribution vs. length of grid edge to satisfy different connectivity levels.

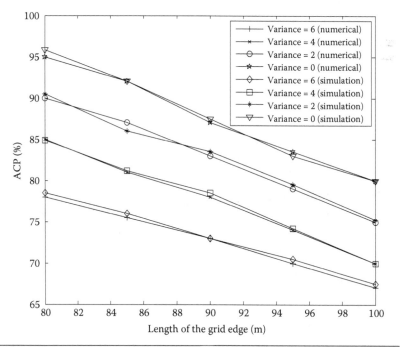

Figure 4.12 Average connectivity percentage vs. length of the grid edge for different error variances.

variance σ^2 increases, the grid ACP decreases, which supports our observation in Figure 4.11. Also, it assures the undesirable effects of placement errors on connectivity as we claimed earlier in this research. The ACP of the sensing field is a monotonically decreasing function of L and this characteristic can be also used to describe the maximum L for a given σ^2 and a given ACP requirement.

4.6 Conclusion

In this chapter, we characterize properties of the grid connectivity and analyze its behavior in real life scenarios, where the node placement is subject to human/machinery mistakes and the communication range is subject to unexpected circumstances that may affect its shape in the 3D space. Thus, the key sources of connectivity failures and problems are addressed. To this end, we propose a generic approach to quantify the average connectivity of the 3D grid under practical random errors in placement and communication.

This approach is applicable to a multiplicity of random error distributions and different grid shapes. We applied the generic approach to normal and uniform distributions of errors in placement with the assumption of an arbitrary 3D shape for the communication range. Numerical results obtained from the proposed approach demonstrate how resilient grid-deployment is to random placement errors and unpredictable (probabilistic) communication ranges. We also demonstrate how to use this research in order to analyze and describe some practical planning issues, and therefore provide a solid guideline for practical grid-based deployment efforts. We can find the maximum allowable placement uncertainty for a given grid-edge length and a given connectivity requirement. Conversely, we can address another practical problem concerning finding the maximum grid-edge length for a given placement uncertainty in order to satisfy a connectivity requirement.

This research can efficiently facilitate the deployment design for realistic connectivity-oriented wireless sensor networks as well as complementing other research efforts made to overcome connectivity problems after their occurrence. We characterize the properties of connectivity and analyze its behavior in real-life scenarios, where the node placement is subject to human/machinery mistakes and the communication range is subject to unexpected circumstances that may affect its shape in the 3D space. Thus, the key sources of connectivity failures and problems are addressed. It has been shown that the grid-based deployment is resilient against the key sources of connectivity problems. We therefore advocate adopting this deployment method whenever possible in practice.

Future work could investigate the resilience of the grid connectivity under other hindering conditions, such as node failures that could occur due to physical damage or limited energy. Another promising direction is studying the effects of temporary hindering conditions, such as mobile obstacles, on the grid connectivity. Also of practical interest is investigating the properties of grid deployment under varying node transmission ranges and/or energy levels.

Acknowledgment

This research is funded by a grant from the Ontario Ministry of Economic Development and Innovation under the Ontario Research Fund-Research Excellence (ORF-RE) program.

References

1. G. Moro and G. Monti, "W-Grid: A Scalable and Efficient Self-Organizing Infrastructure for Multidimensional Data Management, Querying and Routing in Wireless Data-Centric Sensor Networks," *Journal of Network and Computer Applications*, vol. 35, no. 4, pp. 1218–1234, 2012.
2. F. Al-Turjman, H. Hassanein, and M. Ibnkahla, "Efficient Deployment of Wireless Sensor Networks Targeting Environment Monitoring Applications," *Elsevier Computer Communications*, 2011.
3. F. Al-Turjman, H. Hassanein, and M. Ibnkahla, "Connectivity Optimization for Wireless Sensor Networks Applied to Forest Monitoring," in *Proc. of the IEEE International Conf. on Communications (ICC)*, Dresden, Germany, 2009, pp. AHSN11.5.1–5.
4. H. Bai, M. Atiquzzaman, and D. Lilija, "Wireless Sensor Network for Aircraft Health Monitoring," *China Communications*, vol. 2, no. 1, Feb. 2005, pp. 70–77.
5. Y. Jahir, M. Atiquzzaman, H. Refai, and P. G. LoPresti, "Multipath Hybrid Ad Hoc Networks for Avionics Applications in Disaster Area," *28th DASC*, Orlando, FL, Oct. 2009, pp. 23–29.

6. M. Younis and K. Akkaya, "Strategies and Techniques for Node Placement in Wireless Sensor Networks: A Survey," *Elsevier Ad Hoc Network Journal,* vol. 6, no. 4, pp. 621–655, June 2008.

7. Q. Wang, K. Xu, G. Takahara, and H. Hassanein, "Device Placement for Heterogeneous Wireless Sensor Networks: Minimum Cost with Lifetime Constraints," *IEEE Transactions on Wireless Communications,* vol. 6, no. 7, pp. 2444–2453, 2007.

8. F. Wang, M. Thai, and D. Du, "On the Construction of 2-Connected Virtual Backbone in Wireless Networks," *IEEE Transactions on Wireless Communications,* vol. 8, no. 3, pp. 1230–1237, 2009.

9. D. Kotz, C. Newport, and C. Elliott, "The Mistaken Axioms of Wireless-Network Research," *Dept. of Computer Science, Dartmouth College,* Tech. Rep. TR2003-467, July 2003.

10. M. Hashim and S. Stavrou, "Measurements and Modeling of Wind Influence on Radiowave Propagation through Vegetation," *IEEE Transactions on Wireless Communications,* vol. 5, no. 5, pp. 1055–1064, 2006.

11. J. Ryoo and H. Kim, "Multi-Sector Multi-Range Control for Self-Organizing Wireless Networks," *Journal of Network and Computer Applications,* vol. 34, no. 6, pp. 1848–1860, Nov. 2011.

12. G. Tolle, J. Polastre, R. Szewczyk, and D. Culler, "A Macroscope in the Redwoods," in *Proc. of the ACM Conf. on Embedded Networked Sensor Systems (SenSys),* San Diego, CA, 2005, pp. 51–63.

13. V. Ravelomanana, "Extremal Properties of Three-Dimensional Sensor Networks with Applications," *IEEE Transactions on Mobile Computing,* vol. 3, no. 3, pp. 246–257, Sept. 2004.

14. J. Tarng, B. Chuang, and P. Liu, "A Relay Node Deployment Method for Disconnected Wireless Sensor Networks: Applied in Indoor Environments," *Journal of Network and Computer Applications,* vol. 32, no. 3, pp. 652–659, May 2009.

15. A. Erman, L. Hoesel, P. Havinga, and J. Wu, "Enabling Mobility in Heterogeneous Wireless Sensor Networks Cooperating with UAVs for Mission-Critical Management," *IEEE Transactions on Wireless Communications,* vol. 15, no. 6, pp. 38–46, 2008.

16. P. Kulkarni and Y. Ozturk, "mPHASiS: Mobile Patient Healthcare and Sensor Information System," *Journal of Network and Computer Applications,* vol. 34, no. 1, pp. 402–417, Jan. 2011.

17. N. Tamboli and M. Younis, "Coverage-Aware Connectivity Restoration in Mobile Sensor Networks," *Journal of Network and Computer Applications,* vol. 33, no. 4, pp. 363–374, July 2010.

18. A. Abbasi, M. Younis, and K. Akkaya, "Movement-Assisted Connectivity Restoration in Wireless Sensor and Actor Networks," *IEEE Transactions on Parallel Distributed Systems,* vol. 20, no. 9, pp. 1366–1379, 2009.

19. P. Bellavista, A. Corradi, and C. Giannelli, "Mobility-Aware Middleware for Self-Organizing Heterogeneous Networks with Multihop Multipath Connectivity," *IEEE Transactions on Wireless Communications,* vol. 15, no. 6, pp. 22–30, 2009.

20. S. Ross, *Introduction to Probability Models*, 8th ed., AP, London, 2010.
21. D. Seggern, ed., *CRC Standard Curves and Surfaces with Mathematica*, 2nd ed., CRC Press, Boca Raton, FL, 2007.
22. G. Takahara, K. Xu, and H. Hassanein, "Efficient Coverage Planning for Grid-Based Wireless Sensor Networks," in *Proc. of the IEEE International Conf. on Communications (ICC)*, Glasgow, Scotland, 2007, pp. 3522–3526.
23. J. Rodrigues, S. Fraiha, H. Gomes, G. Cavalcante, A. de Freitas, and G. de Carvalho, "Channel Propagation Model for Mobile Network Project in Densely Arboreous Environments," *Journal of Microwaves and Optoelectronics*, vol. 6, no. 1, p. 236, 2007.

5

A DATA DELIVERY FRAMEWORK FOR COGNITIVE INFORMATION-CENTRIC SENSOR NETWORKS IN SMART OUTDOOR MONITORING*

GAYATHRI SINGH AND FADI AL-TURJMAN

5.1 Introduction

Wireless sensor network (WSN) applications have evolved from catering to application-specific requirements to supporting large-scale application platforms such as Smart Cities and Smart Outdoor Monitoring (SOM) in public sensing [1]. These applications typically require a large scale, dense deployment of the sensor network, which generates a large amount of data. However, end users may be interested in accessing specific information from the network (such as temperature in the northeast region of deployment, or pollen alerts for people with allergies). These "smart" application platforms require the underlying WSN to not only gather information from the relevant information sources, but also to prioritize and efficiently manage the heterogeneous traffic flows generated by the requests and deliver information with quality that satisfies the end user's requirements in

* This article was originally published in *Computer Communications*. G.T. Singh, F. Al-Turjman, A data delivery framework for cognitive information-centric sensor networks in smart outdoor monitoring, vol. 74, pp. 38–51, 2016. Reprinted with permission; and also in *IEEE Internet of Things Journal*. G.T. Singh, F. Al-Turjman, Learning data delivery paths in QoI-aware information-centric sensor networks, vol. 3, no. 4, pp. 572–580, 2016. Reprinted with permission.

terms of attributes such as reliability and latency. Providing a good quality of experience to end users in such large-scale deployments requires a shift in focus from traditional address-centric communication abstractions to data-centric routing and storage, where information from multiple, concurrent information sources produced anywhere in the network can be coherently delivered to the end user.

Information-centric network (ICN) is one such paradigm that focuses on content delivery, rather than the point-to-point information flow in the network [2,3]. It makes use of "named data objects" instead of IP addresses to gather data, thus decoupling information source from its location or node identification. ICN is touted as the future technology for content delivery over the Internet because of its ability to bring information to the network layer to improve communication efficiency. Moreover, using the information-centric approach in such resource-rich, static environments, positively impacts data delivery to the end user. Data-centric sensor networks (DCSNs) [4–8] are a parallel paradigm in WSNs where attribute–value pairs are used for named identification of sensed data. Although DCSNs existed much before ICNs, the limited resources and energy capabilities of sensor nodes, and their inability to adapt data delivery decisions to the dynamic network conditions decreased the popularity of this approach in WSNs. Later, with the introduction of the ZigBee standard [9], most of the data processing and communication tasks were off-loaded to relay nodes. However, this also led to a shift to a more address-centric approach for WSNs. Then, with the need to enhance the multiobjective optimization and dynamic decision making capabilities of the network, there was increased research activity in the field of applying cognition to sensor networks. These cognitive sensor networks were able to achieve various goals, such as making the sensor network aware of user requirements; reducing network resource consumption; and making the network exhibit self-configuration, self-healing, and self-optimization properties [10–12]. Despite these advances, it still remains a challenge for sensor networks to differentiate traffic flows in smart environments where the user requirements change over time. Sensor networks still lack the ability to adapt data delivery techniques to different traffic flows generated by the network. In addition, it is desirable to have the sensor network functioning as an information gathering network, to make it easier for users to make name-based requests, and for ease of adaptability to the future ICN.

To cater to all these requirements, we put together the idea of an information-centric approach from ICNs/DCSNs, along with the concept of cognition in this chapter, and propose a cognitive information-centric sensor network (ICSN) framework—COGNICENSE. The information-centric strategy is used to identify relevant sensed information from the network, and the elements of cognition (i.e., knowledge representation, reasoning, and learning) are implemented at special nodes called local cognitive nodes (LCNs) and global cognitive nodes (GCNs), to enhance their information processing and intuitive decision making capabilities. GCNs interpret the user request for the network, and the LCNs help to identify appropriate return paths for data delivery. Relay nodes participate in information transmission over multiple hops, thus maintaining the network's scalability. End user satisfaction is based on the quality of information (QoI) delivered to the sink [13,14], characterized by the attributes of latency, reliability, and throughput associated with the application specific traffic. Accordingly, we summarize our contributions in this chapter as follows:

1. We propose a framework called COGNICENSE that makes use of elements of cognition and an information-centric approach for data delivery in WSN applications for smart outdoor monitoring (SOM).
2. We investigate three QoI attributes: Latency, reliability, and throughput. Based on simulations considering an IEEE 802.15.4 PHY-MAC model, we identify the parameters that affect these QoI attributes.
3. Using a multicriteria decision-making (reasoning) technique called analytic hierarchy process (AHP), we show how the values of the QoI attributes obtained from the simulations can be used to make decisions about the data delivery path that provides the best value of information at the sink (end user).

The rest of the chapter has been organized as follows: Section 5.2 reviews related work in literature. Section 5.3 provides the system models and problem description. Section 5.4 provides details about the proposed data delivery framework using elements of cognition, that is, knowledge representation and inference. Section 5.5 provides simulation results and discussions, and we conclude the chapter in Section 5.6.

5.2 Related Work

The idea of focusing on information objects rather than the host of the information in communication networks is hardly new. Data-centric sensor networks in the wireless world and the TRIAD project [15] for the Internet, described early forms of information-centric networks that aim to move away from the end-to-end communication paradigm and focus on the content being delivered to the end user. In this section, we review DCSNs, and ICNs with respect to their network and design components, and implementation challenges. We also explore the use of cognition in wireless networks with respect to their ability to enable networks to adapt to changing environment conditions, and cater to end user requirements as they evolve with the applications.

5.2.1 Information-Centric Networks

The information-centric network is an information-oriented communication model proposed for the future Internet, to help with managing the huge amount of IP traffic being exchanged globally. Unlike traditional host-centric networks where data routing requires the establishment of single end-to-end path to the host, ICNs decouple senders and receivers by leveraging in-network caching [16,17] and replication of data. User requests for named data objects are addressed irrespective of the source of the publisher or the content's location. This is facilitated by the use of intermediate nodes, which are in-network devices that process and cache named data objects. Thus named data access, routing of requests and data, and information caching comprise the important features of ICNs, and the intermediate nodes play a very important role in implementing these features. These nodes will need to make smart decisions to coordinate their actions and decisions across the network, and also adapt to services and applications as they evolve. Despite the various ongoing research activities in ICNs, not much work is being done with regards to empowering the intermediate nodes to adapt dynamically to changes in the network and end user behavior, to help them learn and evolve on their own.

5.2.2 Data-Centric Sensor Networks

The DCSN approach is very similar to ICN, in naming the sensed objects and in caching data as it is forwarded to the sink. One of the striking differences between DCSNs and ICNs in terms of the network components is that the DCSN approach considers only two types of devices in the network—sensor nodes and sink—whereas ICNs typically use three types of devices—publishers, subscribers, and intermediate nodes. Some DCSNs do propose choosing sensor nodes as cluster heads and involving them in routing data to the sink [18], but this approach burdens the sensor node in terms of energy, data processing, and memory capacities; and it affects the network lifetime and performance on the whole. What has not been explored much in DCSN is applying the ZigBee network model for DCSNs. ZigBee routers are a better choice in terms of conserving sensor's energy and making routers available for more functions, such as information processing, routing, and data caching. ZigBee topology is a big energy saver in terms of off-loading the burden from sensor nodes. Another aspect that has not been explored much in DCSNs is the ability to deal with heterogeneous traffic flows generated in the network as a result of the different request that the network receives. The request could be event-driven, time-driven, query-driven, or a mix of any of these types [19]. Most DCSNs deal with one type of traffic, typically query-driven traffic. However, the challenge is in enabling the network to deal with all types of requests and provide satisfactory service to the end user while adapting to changing network conditions and application requests at the same time [20]. But just as the case with intermediate nodes in ICNs, routers in DCSNs would be burdened with too many responsibilities if they had to carry out all these functions and are not empowered with techniques to deal with them effectively. Hence we look at the possibility of introducing cognition in the routers of the DCSNs.

5.2.3 Cognition in Communication Networks and Cognitive Sensor Networks

To understand the correlation between cognition and communication networks, we'll start with the way wired and wireless communication network architectures have been standardized: The layered protocol stacks of the OSI and TCP-IP models, and the 802 series specifications. As network sizes grew, it became challenging to correlate information from different parts of the network, and to make decisions with incomplete or inconsistent information from different layers of the protocol stack. So the concept of a knowledge plane was proposed by Clark et al. [21] for the wired world, to break the barriers of the layered architecture and enable seamless communication across the layers of the protocol stack and across the network. This idea from the wired world was adopted into wireless networks by Thomas et al. [22], who proposed the idea of a cognitive network. This network would be aware of the application requirements as well as the network dynamics, and make use of learning, reasoning, and feedback from past interactions to make decisions that improve both network performance and end user satisfaction. The feedback in the network is based on an observe–analyze–decide–act loop [23], which, when combined with learning and reasoning, constituted the idea of cognition in the cognitive network. This concept of cognition has been extended to WSNs as well [24], which we will collectively refer to as *cognitive sensor networks* (CSNs) in this work. But these architectures and applications are address-centric, which cater to the end-to-end communication paradigm. To the authors' best knowledge, information-centric architectures (ICNs and DCSNs) have not leveraged the idea of cognition in the way we have described previously to handle diverse traffic flows and satisfy end user requirements simultaneously. Specifically, cognition in data-centric sensor networks can provide the following benefits: (1) In-network information processing (aggregation) can save the energy expended on the huge amount of data exchanged within the network before being delivered to the sink, and (2) using intermediate nodes that incorporate cognition can reduce the burden on sensor nodes and make smart data delivery decisions based on evolving application requirements, and changing environment conditions. Table 5.1 shows a comparison of the infrastructure and data-delivery techniques used in DCSNs, ICNs, and CSNs.

To this end, the COGNICENSE framework we propose will be able to deal with changing application requirements and make smart decisions to provide the requested information to end users with quality that satisfies the SOM application requirements. SOM applications are challenging to handle in terms of the large amount of data

Table 5.1 Comparison of Infrastructure and Data Delivery Techniques in DCSNs, ICNs, and CSNs

	DCSN	ICN	CSN
Network Components	Sensor nodes (SNs) and sink node(s). SNs participate in sensing, transmission, and even data aggregation when they function as cluster heads. Sink nodes disseminate request, store data returned from network, process stored data to respond to user queries, and manage network topology.	Publisher, Subscriber, and Intermediate nodes. Publishers only publish the information. Intermediate nodes deliver published information to the Subscriber. Senders and receivers are decoupled.	Typically address-centric sensor networks with sensor nodes, relay nodes (RNs), and a sink node or base station. In ZigBee-based networks, SNs gather sensed data, transmit to RNs only. RNs participate in multi-hop transmission to sink. Intelligent agents modelled as software agents within network nodes.
Node Deployment and Control	Typically self-organizing. SNs randomly deployed. Dynamic network with centralized control and decision making at the sink.	The ICN environment is a static, resource rich environment for wired communication networks.	Random, deterministic or mixed deployment for network nodes in a dynamic network environment. Distributed control through intelligent agents within the network.
Request Dissemination	Requests are sent out in attribute–value pairs from the sink, which are disseminated in the network through flooding, multicasting, or geocasting, or some combination of multicasting and flooding.	Name-resolution (content name is resolved into components to identify locators for request), or name-based routing (request forwarded based on identifier name).	Request dissemination is mostly address-centric, containing node addresses or end point ids from where data is to be fetched, for end-to-end communication.
Data Gathering/Aggregation	Typically along reverse paths of memorized links, established during request dissemination through broadcast trees; using chains of reporting sensor nodes or through token circulation among equally probable next hop nodes. Data may or may not be aggregated depending on correlation of observed data. Minimum spanning trees are constructed for aggregating data at specific nodes before forwarding them for reporting.	ICNs explore prefix aggregation, request aggregation and aggregation of routing information for functions such as load balancing, and better routing scalability.	Most implementations of CSNs do not depend on or focus on data aggregation methods, or the benefits it can offer. However, data may be aggregated in dense deployments. The cognitive agents focus more on achieving various objectives such as reduced resource consumption, enabling self-organizing and self-healing capabilities of the network and QoS routing under diverse application scenarios.

(Continued)

Table 5.1 (Continued) Comparison of Infrastructure and Data Delivery Techniques in DCSNs, ICNs, and CSNs

	DCSN	ICN	CSN
Cache Storage and Replacement	Information sensed from a given region may vary over time. Hence, stored data may become stale and provide inaccurate information to users demanding current information. Hence, responding to query requires awareness of its type in order to generate useful responses from the network. This traffic classification, and cache replacement policies suitable for such environments do not currently exist.	Caching is inherent in the architecture. Published data doesn't vary over time. Hence, cached information can be reused any number of times and improves network performance over time, as data becomes available from caches closer to the subscriber than the original publisher.	Data storage aspects have not been explored by intelligent agents of CSNs.
Scalability	Scalability and communication range are limited by the use of resource constrained sensor nodes in the network.	The information-centric approach has been proposed to overcome the limitation imposed by IP addressing, for improved scalability.	Since CSNs are based mostly on ZigBee-based communication, scalability is not an issue. RNs provide multi-hop communication over long distances.
Limitations/ Challenges	Energy consumption and delay involved in data processing, aggregation, and delivery. Resource limitations at sensor nodes hinder implementation of advanced routing algorithms and limit caching.	Privacy issues, scalability in caching, cost efficiency.	Cognition has not been explored in a way that can be applied to sensor networks at an architectural level. Implementations are very application/goal specific.

that needs to be handled in-network, and the network nodes are prone to disruptions caused by loss of nodes or poor link quality among communicating nodes [25–27]. Hence, the ability to provide information with QoI attributes of high reliability, low latency, and good hop-to-hop throughput are essential for improving the experience of an end user receiving such data. We make use of an information-centric approach to deal the large amount of information available in the network. Sensed data is identified using attribute tags at sensor nodes. Request for sensory information issued at the sink is routed towards the location(s) in the network where the information has been published. As the request traverses through the network, intermediate nodes are checked for cached copies. As soon as an instance of the desired sensory information is found, it is returned to the sink using cognitive data delivery techniques based on the relative priorities of the QoI attributes that satisfy end user requirements for a given traffic flow.

5.3 System Models

In this section, we explain the COGNICENSE system models and its core components in details, in addition to listing our main assumptions.

5.3.1 Quality of Information (QoI)

QoI is defined as the level of satisfaction experienced/perceived by the end user on the information received from the network [13]. Attributes such as reliability, latency, and throughput are used to evaluate the QoI of data delivered to the sink. To differentiate QoI from quality of service (QoS) of WSNs [28], QoS takes care of the operational aspects of the network, while QoI is associated with the characteristics of the sensory information made available to the end user. In our proposed approach, priorities are evaluated for these QoI attributes for each application traffic type at the sink, and the network tries to deliver the information with the desired QoI to the sink/end user. For SOM applications in WSNs, QoI attributes that help us assess how well the network is able to gather and provide relevant sensory information is based on the following QoI attributes: Reliability, latency, and throughput. Their definitions are based on the work in [29], and are presented here briefly:

Latency (L): Defined in terms of the mean frame service time at the MAC layer and is estimated as the time interval from the instant a packet is at the head of its MAC queue and ready to transmit, until an ACK for such a packet is received. In other words, it is the average delay for a successfully received packet.

Reliability (R): Defined as the probability that a frame is not blocked, lost due to channel access failure, or discarded as a result of reaching the maximum number of retries limit.

Average throughput (AT): Is a function of reliability and is defined as λ * Reliability * Application load (bits), where λ is the average frame arrival rate at a node in bits/second.

Instantaneous throughput (IT): Is a function of latency and is defined as application payload (bits)/Latency(s). We use the instantaneous throughput value for computations in our work, and refer to it simply as "T."

5.3.2 Network Lifetime

In this work, we propose a novel definition for network lifetime based on the quality of information (QoI) perceived by the end user. Network lifetime is defined as "The time or number of transmission rounds beyond which the network can no longer deliver useful information to the end user. This is reflected by the network's inability to find a data delivery path with satisfactory values for QoI attributes (latency, reliability, and throughput), as determined by the end user, or when there is insufficient energy in the network nodes to deliver such data to the sink for any of the application generated requests."

This definition not only caters to satisfying the application requirements, but also considers the status of the network and node resources (especially in terms of remaining energy at the nodes) in defining the network lifetime. If sensor nodes or LCNs were drained of energy, then at each hop, the QoI attribute values would be affected, and thus reflected in the overall value of information delivered at the sink. Thus it also justifies the fact that if the network doesn't have sufficient resources to deliver data, it cannot satisfy the end user, and hence it should be considered as the end-of-life of the network, as no useful information can be derived from it.

5.3.2.1 Application Traffic Profiles for Smart Outdoor Monitoring Applications Application traffic is profiled into three categories [30] based on how often sensed information from the network needs to be delivered to the end user, and the priorities associated with the QoI attributes for each traffic type. They traffic profiles are as follows:

Type I: Periodic (application defined rate).
Type II: Intermittent (application/external stimulus defined rate) or event driven/query driven traffic.
Type III: Low-latency data (emergency/alerting information).

We illustrate this traffic classification by making use of a sensor network deployed in the following SOM applications. The first one is a sensor network deployed for urban environment monitoring. In this application, traffic flow for an air-quality monitoring station is classified as Type I. Information flow generated in response to queries from an operator or end user, requesting for specific information such as temperature or humidity at a specific time of the day is classified as Type II traffic. Finally, a service that issues alerts, such as high ultraviolet radiation warning, heat wave warning during extreme temperatures, reduced visibility warning, and pollen alerts, has traffic flow corresponding to Type III.

Another example of a SOM is a sensor network deployed for monitoring a forest environment [31]. When the network transmits information corresponding to periodically sensed data from the forest region, the flow corresponds to Type I traffic. Information flow corresponding to the assessment of factors that influence the type of flora and fauna found in the monitored region is classified as Type II traffic, and traffic flow associated with alerts issued in emergency situations such as forest fires is classified as Type III traffic.

5.3.2.2 Network Architecture and Components Figure 5.1 represents the components of the COGNICENSE framework and their interactions. Sensor nodes (SNs), relay nodes (RNs), local cognitive nodes (LCNs), and global cognitive nodes (GCNs) constitute the nodes of the cognitive information–centric sensor network (CICSN). SNs

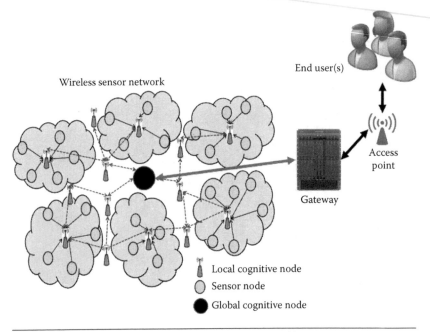

Figure 5.1   The cognitive information-centric sensor network architecture.

constitute the leaf nodes that are deployed uniformly and randomly in the network. They communicate with LCNs and RNs lying within their communication range. Typically, SNs communicate with only one parent LCN or RN at a time. LCNs communicate with each other, with RNs, and a cognitive sink node called the GCN, which is located at the center of the deployment region. The GCN carries information to and from the sensor network to the end user through a gateway and access-point. When hierarchically represented, the CICSN node interactions are as depicted in Figure 5.2(a). LCNs and RNs are deployed at predetermined locations on a grid as shown in Figure 5.2(b), so as to ensure complete coverage of the target area and connectivity of SNs with the GCN.

5.3.2.2.1 Cognition in ICSNs   Haykin [32] and Mitola [33] have perhaps defined cognition in its most extensive form in the context of wireless communication systems. Going beyond simple adaptations, they make use of a feedback loop: The observe–analyze–decide–act (OADA) loop [20], to model cognition in a way that doesn't deal with imitating human-like behavior, but in making intuitive decisions

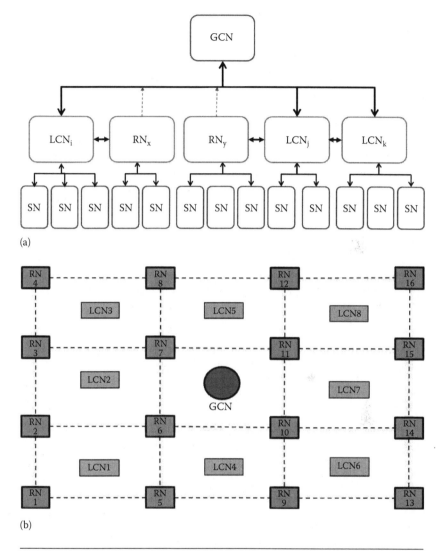

Figure 5.2 (a) Hierarchical organization of network nodes in the CICSN. (b) Representation of LCNs and RNs in a two-dimensional grid structure.

based on learning from the environment to adapt to current network conditions while inferring from past behavior and knowledge to predict a course of action for the future that the network can benefit from.

Based on this idea, and drawing from the work on cognitive networks [34] and extending our work on cognitive information-centric sensor networks [35,36], we define *elements of cognition* to implement the functionality of the OADA loop. Knowledge representation, reasoning, and learning constitute the elements of cognition, which when implemented in specialized nodes of the network, will help them make cognitive decisions, and make the sensor network a cognitive one. In the CICSN, LCNs and GCNs are the specialized nodes that implement the elements of cognition.

5.3.2.3 Node Functions   In this section we describe the functions of the sensor, relay, and cognitive nodes of the ICSN. We start with the sensor nodes. Sensor nodes host a multitude of sensors as required by the application platform. Raw sensed–data is stored in attribute–value pairs. This representation facilitates named-data identification to locate the user-requested information. Thus, the two main functions of the sensor nodes are (1) sensing raw-data and (2) storing sensed information in attribute–value pairs. Details of the attribute–value pair representation follow in Section 5.4.1, where we deal with Knowledge Representation. They communicate with relay and local cognitive nodes. Relay nodes communicate with SNs and LCNs to act as intermediate nodes that gather information from SNs, and forward them to their LCN neighbors. They deliver data over multi-hop paths to the GCN.

LCNs perform two main functions: (1) Gathering sensory data from sensor nodes and forwarding information from relay nodes, and (2) data delivery based on QoI requirements of the traffic type. LCNs also function as caches to store the data as it travels through the network. LCNs make use of the sensor attributes to identify the relevant data, similar to the named data–object search in ICNs and DCSNs. The requirements on the QoI attributes are based on the type of traffic flow generated as a result of the end user's request. As for dealing with the QoI attribute requirements, an analytic hierarchy process (AHP) [14,37] is implemented as the reasoning element of cognition to make the decision in the LCNs.

GCNs have the following main functions. They receive user requests and synthesize them to identify the following information: Application traffic type, requested sensor attributes, and QoI attribute priorities. They broadcast the synthesized information to the LCNs, so that they may process it further to gather the requested information from the network. Once the network returns the requested information, GCNs process it to determine if the QoI provided by the network meets with the user requirements, and deliver information with acceptable QoI to end user. They also determine when the network is no longer able to deliver useful information from the network, thus flagging the end-of-life of the network.

5.4 The COGNICENSE Framework

Elements of cognition in the network nodes and an information-centric data delivery approach are the two main constituents of the COGNICENSE framework. The elements that help in implementing cognition in the cognitive nodes are knowledge representation, reasoning, and learning. Knowledge representation helps in identifying data using attribute–value pairs and contributing towards identifying named-data objects for the information-centric approach. Reasoning helps in multicriteria decision making to prioritize the QoI attributes for a given traffic flow, and decide on the number of sensor nodes chosen for data transmission to the LCN, or the next hop node chosen along the data delivery path to the GCN. While reasoning helps in achieving short-term objectives and making decisions that help the current situation, learning helps in achieving long-term goals of the network, such as improving its lifetime. Feedback obtained from the network's past behavior aids the learning process, and helps in planning proactive responses to changes in network behavior and user requests.

5.4.1 Knowledge Representation

A frame structure based on attribute–value pairs is used in sensor nodes and the cognitive nodes for knowledge representation. In frame-based knowledge representation [38], the frame is defined as a hierarchical data-structure with inheritance [39]. It has slots which are function-specific cells for data. In sensor nodes, these function-specific cells store sensor attribute–value pairs. In LCNs, they store more information, such as the one-hop neighbor LCNs and the associated values of QoI attributes in the last communication round. Information accumulated over several rounds of information transmission leads to the formation of a knowledge base (KB), which can be looked up by the reasoning mechanism to make quick decisions on choosing the data delivery path which satisfies the QoI delivered to the end user.

We make use of a semantic naming scheme using strings (sequence of characters) that provide information about the originator of the request, traffic type expected to be generated in response to the request, direction from which the data is requested, and the sensor data attribute(s) corresponding to which the data is to be gathered. The naming scheme has two main components: (1) Request classifier and (2) information attributes. The request classifier (RC) field has two subfields: The originator of the request and the type of traffic expected. The information attribute (IA) component also has two subfields: Direction attribute and sensory data attribute. The two fields are separated by a colon ":" and the subfields within a field are separated by an underscore "_". Here is the format of a request string: <Request_classifier>: <Information_Attribute>. Let us consider an example request string. *Sink_type1:N_temp*. Here, "sink" indicates that the request has been originated by the sink. "Type1" indicates that the expected response from the network is a periodic traffic flow. "N" indicates that the direction from which the data is expected to be gathered is north. "Temp" indicates that temperature data is being requested.

Thus, the request string means: Sink initiated a request to collect periodic data from the Northern region of the deployment for the temperature attribute. Further, a combination of logical and relational operators can be used to add more details in the request. For example, the request string *Sink_type1-60:N_temp&&humd* specifies that temperature and humidity values are to be returned periodically, every 60 minutes. Once a complete match is found for the request string, the data is returned in attribute–value pairs to the sink by concatenating it to the original request string using a ":" operator and changing "Sink" to "Source." For example, the response string *Source_type1:N_temp:temp-25_temp-26_temp-24* indicates the temperature–value pairs recorded were 24°C, 25°C, and 26°C.

The alphabet required for a complete representation of this language are represented in Table 5.2.

For further digitizing the representation, each of the alphabet's values can be uniquely binary encoded. The cognitive nodes (GCN and LCN) will be able to generate and parse these strings and arrange the

Table 5.2 Alphabet Used for Representation of Attributes as Semantic Information

ALPHABET	VALUES	REMARK
α (Information Source)	{Sink, Source}	Indicates if this is a request or response.
β (Traffic Type)	{type1, type2, type3}	Traffic flow type expected in the network in response to request.
γ (Direction attribute)	{N,E,W,S,NE,NW,SE,SW,ALL}	Direction(s) from which data may be requested. "All" indicates broadcast throughout the network.
δ (Attributes of Sensed data)	{temp, humd, uvi, co2,time}	Sensory attributes for which data can be provided by sensor nodes. "Time" indicates the time stamp at which data was registered at the sensor node.
Logical and relational operators	&&, >, <. >=. <=	

information gathered from SNs/RNs in the desired format. The information attribute (IA) component also has two subfields: Direction attribute and sensory data attribute. The two fields are separated by a colon ':' and the subfields within a field are separated by an underscore '_'. Here is the format of a request string:

<Request_classifier>: <Information_Attribute>.

Let us consider an example request string. *Sink_type1:N_temp*. Here, "sink" indicates that the request has been originated by the sink. "Type1" indicates that the expected response from the network is a periodic traffic flow. "N" indicates that the direction from which the data is expected to be gathered is north. "Temp" indicates that temperature data is being requested. Thus the request string means: Sink initiated a request to collect periodic data from the Northern region of the deployment for the temperature attribute. Further, a combination of logical and relational operators can be used to add more details in the request. For example, the request string *Sink_type1-60:N_temp&&humd* specifies that temperature and humidity values are to be returned periodically, every 60 minutes. Once a complete match is found for the request string, the data is returned in attribute–value pairs to the sink by concatenating it to the original request string using a ":" operator, and changing "sink" to "source." For example, the response string: *Source_type1:N_temp:temp-25_temp-26_temp-24* indicates the temperature–value pairs recorded were 24°C, 25°C, and 26°C.

The alphabet required for a complete representation of this language are represented in Table 5.2. For further digitizing the representation, each of the alphabet's values can be uniquely binary encoded. The cognitive nodes (GCN and LCN) will be able to generate and parse these strings and arrange the information gathered from SNs/RNs in the desired format.

5.4.2 Learning

Learning is used in the COGNICENSE framework for identifying data delivery paths towards the GCN that satisfy the user's requirements in terms of QoI attributes. In this work, we make use of a direction-based heuristic to determine the data delivery path through RNs that lie in the direction of the GCN. This means that each time an LCN has to choose from among multiple RNs to decide the next hop, the direction-based heuristic eliminates RNs that increase the distance between the current LCN and GCN. Knowledge of the positions of the LCN and its one-hop RNs is used by the heuristic to determine the set of such RNs, which we call *forward-hop-RNs*. Thus the forward-hop-RNs of an LCN identified by the direction-heuristic is constituted by those RNs that reduce the distance between the LCN and the GCN. This information is stored in the LCN's knowledge base for use in the next transmission rounds. Feedback about QoI delivered along the forward-hop RNs is used to identify the best forward-hop RN for each traffic type. Thus the direction-based heuristic, along with feedback from the network about the QoI delivered along the chosen paths helps the LCNs to learn data delivery paths to the sink, as the network topology changes due to link variations and node deaths.

5.4.3 Reasoning

An analytic hierarchy process (AHP) is used for implementing the reasoning element of cognition. AHP aids with multiple-criteria decision making while deciding on the data delivery path based on the quality of information requirements of the requested application.

For example, for Type III traffic, requesting for low latency data, the QoI requirements are as follows: Highest priority is latency, followed by reliability, and finally throughput. This means that while choosing the next hop node for data delivery, the node that provides the lowest latency will be chosen. Reliability is more important than throughput. Hence, if two next hops guarantee the same latency then the next attribute to compare will be reliability, and lastly, throughput. AHP provides a method for pairwise comparison of each of the QoI attributes and helps to choose the node that can provide the best value of information with respect to all three QoI attributes. Subsequent sections have more details with a running example on AHP. While these calculations help in deciding the next hop, they also help in planning for future actions. The cognitive nodes are able to store the calculated priorities of the QoI attributes, which they can use to decide which type of traffic the LCNs can best provide for. Hence, these calculations need not be necessarily calculated for every transmission round.

5.4.3.1 The AHP Framework for Data Delivery Based on QoI Attributes There are three levels in the AHP hierarchy: Goal, criteria, and alternatives, as shown in Figure 5.3.

- *Goal*: Deliver application-requested sensory information to the GCN from LCN by identifying the next hop node.
- *Criteria*: Data must be delivered with the appropriate priorities of QoI Attributes for each application type. The QoI attributes that are considered are latency, reliability, and throughput.
- *Alternatives*: The RNs in the network available to forward the data over multiple-hops in the network.

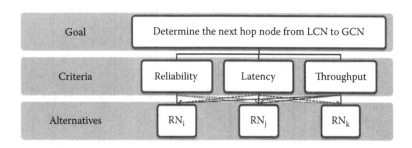

Figure 5.3 The AHP hierarchy.

A fundamental scale for pairwise comparisons is then used to set application-defined priorities for the QoI attributes [37]. Then the priorities of QoI attributes are established using pairwise comparison. Let us consider an example where a SOM application wishes to transmit low-latency alerting information to its users. From the three QoI attributes of latency, reliability, and throughput, we would assign the highest priority to latency, to ensure timely delivery of the alert, followed by reliability, and then throughput. We tabulate the relative priorities of each the QoI attributes using pairwise comparison and generate Table 5.3. Then, the AHP computation involves generating the Eigen vector for the values in Table 5.4, using the following steps:

1. iRepresent the values of Table 5.4 in matrix form {A = [1,4,6;1/4,1,3;1/6,1/3,1]}
2. iCompute the Eigen vector of the matrix A {[v,d] = eig(A)}
3. iIsolate the absolute, real values of the Eigen vector {q=abs(real(v(:,1)))}
4. iCompute the normalized, relative priority values as {Relative_Priorities=q/norm(q,1)}

This way, the AHP algorithm is implemented at LCNs to establish the relative priorities of QoI attributes and helps in multicriteria decision making. The QoI attributes are the criteria and the goal is to find the next hop RN during data delivery from LCN towards GCN, which provides the highest value for a specific QoI attribute, or provides the overall best value of information (VoI) as illustrated in Table 5.5. VoI is based on the combined value of QoI attributes and

Table 5.3   Pairwise Comparison of QoI Attributes

Latency	4	Reliability	1
Latency	6	Throughput	1
Reliability	3	Throughput	1

Table 5.4   AHP for QoI Attributes v/s Goal

GOAL: BEST QoI	LATENCY	RELIABILITY	THROUGHPUT	RELATIVE PRIORITIES OF THE QoI ATTRIBUTES
Latency	1	4	6	0.691
Reliability	1/4	1	3	0.2176
Throughput	1/6	1/3	1	0.0914

Table 5.5 AHP to Evaluate the Overall Priorities for All Possible Next-Hop RNs

BEST CANDIDATE FOR NEXT HOP RNX	PRIORITY WITH RESPECT TO			
	LATENCY	RELIABILITY	THROUGHPUT	GOAL
RNi	**0.252**	0.015	0.101	**0.375**
RNj	0.2	0.018	0.11	0.329
RNk	0.164	**0.019**	**0.116**	0.296

energy consumed during the process of delivering information to the GCN.

VoI delivered to the end user is said to be maximized when data is delivered over links that provide the best effective QoI for each traffic type, while minimizing the energy consumed in the network while doing so.

$$Value\ of\ Information\ (VoI) = \sum_{n-hops} (Effective\ QoI) - \sum_{n-hops} (Energy\ Cost)$$

(5.1)

Equation 5.1 highlights that lower the energy cost of delivering data to the sink, higher is the VoI associated with that data/information object. The QoI must be maximized and energy cost minimized to achieve the best VoI. If energy consumption is measured as a function of the number of transactions taking place before data is delivered to the GCN, a simple metric—the hop count—can be used to approximate the energy cost. If the information is transmitted from source to GCN over minimum number of hops, each link providing the best combined QoI for that traffic type, we can say that the information was delivered to the GCN with good VoI. The steps used in the AHP to establish priorities for the QoI attributes and identify the best next hop path in delivering the application data to the GCN are illustrated in Algorithm 5.1.

Information about the relative priorities of the QoI attributes as desired by the user are received as input from GCN in steps 1–3. The output is a next hop RN that provides the best QoI as shown by steps 4–5. The simulations are set to run till no path can be found to GCN or till 50 percent of RNs and LCNs die. In steps 9–11, AHP analysis identifies the best next hop RN that satisfies these requirements, and identifies the next hop path for data transmission. Steps 12–17 define actions to be taken when data reaches the GCN and leads to a successful transmission, or reaches another LCN from where next hop has to be identified. Steps 18–21 indicate that if a path to GCN was not found along the chosen path, GCN issues a retransmit request. These computations can be initially carried out for each next hop node decision in the data delivery path. This technique helps to build the learning database at each LCN about its next hop neighbor, and the priorities each of them offers with respect to the QoI attributes. This information can be stored and used for planning future rounds of data delivery for application traffic that may need to choose a different next hop for the same source LCN, based on the expected values of attribute priorities at the GCN. Thus we can see that this AHP process helps in adaptive multicriteria decision making during data delivery, in considering the desired attribute priorities for each application-traffic type.

Algorithm 5.1 AHP Analysis to Determine the Data Delivery Path

1. **Function AHP (QoI.priorities)**
2. **Input**
3. **QoI.priorities:** End user defined priorities on QoI attributes for requested data
4. **Output**
5. **RN_x:** Forward-hop $RN_x \in \{RN_1 ..... RN_n\}$ with best QoI
6. **Begin**
7. **Initialize:** QoI priority matrix for traffic type; Success=0;
8. **While** (*number of dead nodes<50 percent or network not disconnected*)
9.   *AHP_analysis*(Next hop RNs v/s QoI attributes)//
10.   Next hop RN = RN_x//This is the RN with best QoI for chosen traffic type
11.   Transmit data to next hop RN
12.   **If** (next hop = GCN)
13.   Success=1;
14.   **Else**
15.   *Choose next hop LCN*
16.   goto step 8

17.   **End**
18.   **If** *(Success==0)*
19.   GCN Retransmits request
20.   **End**
21. **End**

5.4.3.2 Node Mobility Support in the COGNICENSE Framework The COGNICENSE framework allows for caching data at LCNs that act as intermediate nodes. This makes data readily available for users at nodes other than sensor nodes, thus offering two main advantages: (1) It prevents requests being sent out to sensor nodes, which may be in a sleep cycle, leading to a lost request and (2) it helps to conserve valuable energy resources by reducing the number of transmissions occurring in the network; both from sensor nodes towards the sink, and over multiple relay nodes that transmit the sensory information from the sensor nodes to the sink. Furthermore, the named-data identification enhances the advantages offered by the data caching feature at the LCNs in terms of supporting node mobility. We discuss the issue of node mobility under two categories: (1) Sensor node mobility, and (2) LCN mobility.

5.4.3.2.1 Sensor Node Mobility Support In the COGNICENSE framework, search for data is name-based, which means that the request is not associated with any specific address, location or an end point. This is in contrast with the IP based approach, where an address is associated with each sensor node, and the request–response cycle involves the establishment and maintenance of an end-to-end connection between the sensor node and the sink. This restricts the ability of the network to support node mobility, as the loss of connectivity with the source–sensor node or any intermediate node involved in the end-to-end connection, due to node death or lossy links, affects the data gathering and routing capability of the network.

However, in a cognitive ICSN, the requested information could be located anywhere in the network, and the user will be able to access it, since the request is not tied with any specific node address. Any node that can provide a match to the requested information can provide the data. Moreover, the routing path is not fixed, and can adapt to the changing network topology. This is made possible by the LCNs that make use of cognitive reasoning to dynamically identify data delivery paths based on the type of request, and how well a link had performed in a previous round. The data delivery paths are chosen based on the QoI attributes of latency, reliability and throughput. The LCNs offer another advantage of acting as a data cache. Information gathered from sensor nodes can be stored in these LCNs to make them available on-demand, without having to access the source sensor nodes. We assume that cooperative caching techniques designed for wireless sensor networks [40] that deal with large amounts of sensed data, can be applied at the LCNs to enable them to manage information storage. In addition, we assume that cache replacement algorithms such as least value first (LVF) replacement [16] can be used to maintain availability of relevant data while evicting stale and unused data from the cache, to make space for fresh data. Thus, even if a source sensor node was mobile, the sensed information is stored in LCNs whenever the node lies in close proximity with the LCN, and is made available to the user, irrespective of the mobility condition and/or pattern of the sensor node. Thus the COGNICENSE framework is capable of supporting sensor node mobility, without negatively affecting the network performance.

5.4.3.2.2 Local Cognitive Node mobility support   A further advancement that can be made to the COGNICENSE framework, is the ability to support LCN mobility. A combination of static and mobile data collector LCNs could be used in the information-centric sensor network to improve the data gathering capability of the network. The advantage offered by having mobile LCNs is that, when a part of the network starts to deteriorate in its energy capacity and link conditions, the mobile LCNs will still be able to gather information from that part of the network, and store it in their cache. Thus preventing a part of the network from getting completely disconnected from the rest of the network, as long as the sensor nodes remain functional. These mobile LCNs could then communicate amongst themselves and with the static LCNs, to decide on the best way to deliver the collected data to the sink, and also to maintain information about the entire network to make informed decisions while responding to user requests.

5.4.3.3 Energy Considerations in the COGNICENSE Framework Energy conservation is one of the most important aspects of WSN design. In ZigBee based address-centric WSNs, sensor nodes off-load the energy-draining communication tasks to relay nodes. SNs being leaf nodes do not have the network layer to forward data beyond their one-hop relay nodes, and they do not even communicate amongst each other. The multi-hop relaying between source and sink is done by RNs, which have higher battery and processing capacity. Let us denote the energy cost of the relay node using Equation 5.2:

$$C_{RN-E} = C(TE_{Tx} + RE_{Rx}) \tag{5.2}$$

Most of the energy consumption at the RN is due to data communication, represented by E_{Tx} for energy consumed during transmission and E_{Rx} for energy consumed during data reception. T and R represent the number of transmitted and received packets respectively. Let us compare this energy with that at the cognitive node (C_{CN-E}). Typical functions of CNs that consume additional energy compared to regular RNs are data aggregation and the cognitive decision process. Additional energy consumption is accounted for by two factors: (1) Protocol overhead incurred during cognitive data delivery due to feedback from the network during the learning process and the exchange of values of QoI attributes such as latency, reliability, and throughput while making routing decisions and (2) increased transmit power for increasing the communication range of CNs.

$$C_{CN-E} = C(TE_{Tx} + RE_{Rx}) + C(AE_{ag}) + C(PE_{cog-process}) \tag{5.3}$$

In Equation 5.3, $T, R, A,$ and P are the total number of packets that are transmitted, received, aggregated, and processed by the cognitive elements respectively in each transmission round. $C(TE_{Tx} + RE_{Rx})$ is the energy cost incurred during data transmission and reception, $C(AE_{ag})$ represents the energy cost incurred during data aggregation and $C(PE_{cog-process})$ represents the energy cost due to protocol and processing overhead during the cognitive processes. Expressing Equation 5.3 in terms of the energy cost of RNs we get:

$$C_{CN-E} \geq C_{RN-E} + AE_{ag} + CE_{cog-process} \tag{5.4}$$

If the relay and cognitive nodes use the same transmit power, then the equality sign holds true in Equation 5.4. In any case, the energy cost of the cognitive node is higher than that of the relay node. In order to ensure that the energy cost of CNs does not offset the advantages offered by it in terms of adapting to the dynamic traffic flows and network topology changes, the cost can be optimized by maximizing the number of RNs and minimizing the LCNs in the deployment.

5.5 Simulations and Results

A CICSN for a SOM application was implemented on top of an IEEE 802.15.4 MAC-PHY simulator [41,42] in MATLAB®. The deployment and interconnection among the network nodes (SNs, RNs, LCNs, and the GCN) is as shown in Figure 5.4. The cyan and

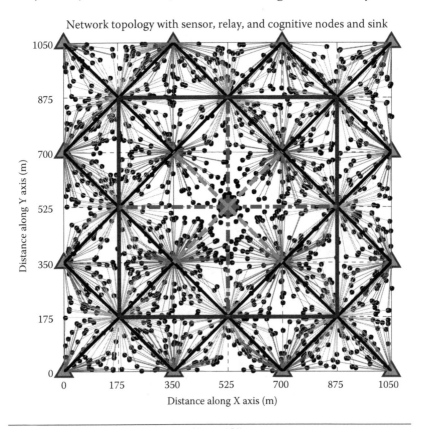

Figure 5.4 Interconnection among SNs, LCNs, and RNs.

magenta lines indicate links between SNs and LCNs and SNs and RNs, respectively. GCN in red is located at the center of the target area. Blue lines show inter-LCN communication links and the black lines indicate interactions between LCNs and RNs. Green lines indicate the links between the GCN and its one-hop RNs, and the red lines are the links between the GCN and nearest LCNs. The simulations were used to evaluate the impact of network and node parameters on QoI attributes.

Using parameters identified from this simulation, the AHP based data delivery technique (AHPDD) was implemented, and its performance was compared with two other techniques—a multipath data delivery technique (MDD), and a higher remaining battery-based data delivery technique (HRBDD), in terms of the number of data transmissions to the GCN, and the QoI along the data delivery path.

5.5.1 Simulation Setup and Parameters

The first set of simulations was used to identify parameters that affect the QoI attributes of latency, reliability, and throughput, for the application. Parameters chosen for observation were (1) N_active: the number of nodes attempting to simultaneously transmit data, and (2) the offered load: the per-node frame arrival rate expressed as a fraction of the application payload in bits per second. The simulation was setup to identify the impact of varying the offered load on the QoI attributes for different values of N_active. The maximum and minimum possible values for N_active were chosen based on the node binding information obtained from the deployed CICSN. From 10 sets of random deployment of sensor nodes, we found a lower bound of about 10 sensor nodes per LCN and an upper bound of close to 60 sensor nodes per LCN. The range of values for per node offered load was 0 to 1400 bits per second, such that the load could be expressed as a fraction of the application payload, ranging from 0.1 to 1.4 times the size of the maximum application payload of 121 bytes. The remaining simulation parameters were set as shown in Table 5.6.

Table 5.6   Parameters of the Simulated CICSN

PARAMETER	VALUE
Target area	1050 m × 1050 m
Number of nodes	SNs: 1500
	RNs: 16
	LCNs: 8
Transmit power	SN: <3 dB
	RN: 3 dB
	LCN: {3 dB, 5 dB, 7 dB}
Communication range	SN: 175 m
	RN: 250 m
	LCN: 350 m
	GCN: 500 m
Application payload size	121 Bytes
Per node offered load	0–1400 bits/s

5.5.2 Simulations Showing the Impact of Network and Node Parameter Variations on the QoI Attributes

The impact of varying the offered load and N_active on the QoI attributes of latency, reliability, and throughput for the SOM application is analyzed using the simulation results in Figure 5.5. Figure 5.5(a) indicates that latency increases almost linearly with increase in offered load for small values of N_active, up to 10 nodes. However, for higher values of N_active, latency saturated around 0.1 seconds for loads greater than 1000 bps. Figure 5.5(b) shows an overall trend of decrease in reliability as the offered load increases. However, there is a marked difference in the variation of reliability with increase in N_active. Reliability drops exponentially for values of N_active greater than 30, as offered load increases. For values of N_active around 20, reliability remains around 1 for loads up to 500 bps per node, after which it drops linearly with increase in offered load. Figure 5.5(c) indicates an overall decrease in throughput as offered load increases. For N_active = 10, the decrease is linear, but for higher values of N_active, (20 nodes and above), the decrease in throughput with increase in offered load is exponential. Figure 5.5(d) indicates a very different trend compared with instantaneous throughput at N_active = 10. There is an increase in throughput with increase in offered load, and stabilizes at around 700 bps for offered load over 1250 bps. However, as the value of

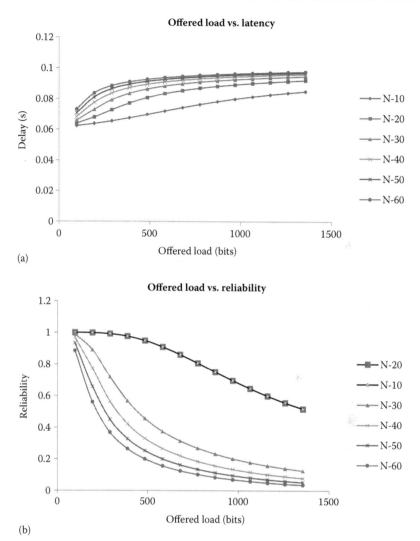

Figure 5.5 (a) Impact of N_active and offered load on Latency. (b) Impact of N_active and offered load on Reliability. (*Continued*)

N_active is increased, the absolute value of throughput decreases, and the increasing trend in throughput that was seen for N_active = 10, starts reversing for loads greater than 500 bps for N_active over 30. We made the following observations from analyzing the impact of varying the per-node offered load and N_active on the QoI attributes: (1) Values of each of the QoI attributes deteriorates as the offered load increases, and (2) restricting the number

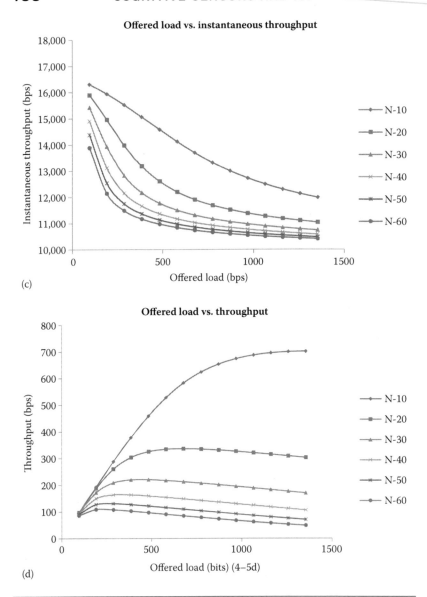

(c)

(d)

Figure 5.5 (Continued)   (c) Impact of N_active and offered load on instantaneous throughput. (d) Impact of N_active and offered load on average throughput.

of nodes attempting to simultaneously transmit data (N_active) to around 10 nodes helps to achieve good values for all the QoI attributes. We use these observations to setup the simulation parameters for our next set of simulations.

5.5.3 Comparative Evaluation of Data Delivery Protocols: AHPDD, HRBDD, and MDD

Using the aforementioned observations, a network environment in which less than 10 nodes are scheduled at a time for simultaneous transmission, and the maximum transmission load is limited to five frames per second (fps) is set up. Channel conditions were varied by varying the application payload and N_active values. We assume that LCNs and RNs start with an initial energy of 25 units, and SNs have an initial energy of 15 units. Each transmission from a SN consumes one unit of its energy; and transmissions from RN to LCN and vice-versa consumes two units of energy at the transmitting node. Direct communication among LCNs or LCN to GCN consumes three units of power. These values are based on the transmit power and communication range capabilities of the nodes. At the start of the simulation, we identify a source LCN at which required data is available. Delivery of data from the identified source LCN to the GCN is considered as one successful transmission round. Using this setup, we analyze the performance of the AHP-based data delivery protocol (AHPDD) based on the number of transmission rounds of delivering data from a source LCN to GCN, until one or both of the following simulation termination conditions are satisfied: (1) 50 percent of the total number of LCNs and RNs die out, or (2) the network is no longer able to deliver information to the GCN as all the one-hop neighbor RNs and LCNs to the GCN are dead. At this point, the simulations are terminated. AHP analysis is implemented at LCNs to identify the best next hop RN. The priority matrix for AHP analysis is set to identify data delivery path for each of the three traffic types. The AHP-based decision protocol is then compared with two other decision criteria in the same network setup, but without considering the cognitive reasoning capabilities at the LCN or GCN. These routing strategies are based on the ones described by Stojmenovic [7] for reporting via alternate paths in a broadcast tree in DCSNs.

The first one is based on choosing an RN with the highest remaining energy from among the one-hop neighbor nodes, and is called highest remaining battery based data delivery technique (HBRDD). The second one is called multipath data delivery (MDD), where each node transmits through all its one-hop neighboring nodes with equal probability to improve the chances of successful data delivery to the sink. Data is delivered via multiple paths at each hop, until at least one of the paths leads to the sink, which is the noncognitive version of the GCN. The simulations were allowed to run till one or both the simulation termination conditions were met, and the average value of 25 such simulations was taken. The number of transmission rounds during which data was not delivered to the GCN was also recorded. The following criteria were used to determine unsuccessful transmissions to the GCN: (1) Inability of the routing protocol to forward data to the GCN due to node deaths along the path chosen for data transmission, and (2) transmission failure due to insufficient remaining energy at the forwarding nodes. The difference between the total number of transmission rounds, and the number of failed transmissions gives a measure of the number of transmission rounds in which data was successfully transmitted to the GCN. Thus we define the failure rate of the routing protocols in Equation 5.5 as follows:

Failure Rate = (*Number of failed transmissions/Total number of transmission rounds*) * 100

$$(5.5)$$

From the simulation results in Figure 5.6, we can see that AHPDD and HRBDD perform equally well, and better than MDD, in terms

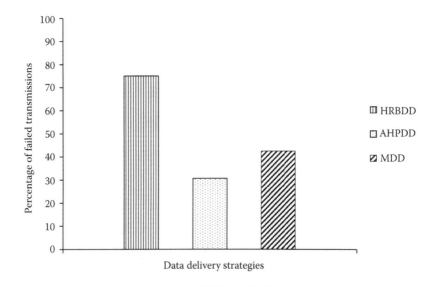

Figure 5.6 Comparison of failure rates.

of the number of transmission rounds. However, from Figure 5.7, we see that the number of failed transmissions is very high for HRBDD (57 out of 76). On comparing the failure rates, we find that MDD in fact performs better than HRBDD by 12 percent. While only 31 percent of the transmissions using AHPDD fail to reach the GCN, the failure rate is as high as 75 percent with HRBDD, which is almost twice as worse when compared with the 42 percent failure rate of MDD. Figure 5.8 shows the number of successful transmission rounds for each of the data delivery techniques. We see that although MDD doesn't keep the network running for more number of transmission rounds compared to HRBDD, it is able to deliver data to the sink successfully for an average of 42 percent of the total transmission rounds, which is 17 percent higher than what is achieved by the HRBDD. However, AHPDD out performs both these protocols by adapting the data delivery decisions to user priorities, and successfully delivering data to the GCN for 70 percent of the total transmission rounds. From these simulations, we can say that AHPDD is better able to adapt to the changing network topology and deliver data to the GCN with a lower failure rate compared to the other two techniques.

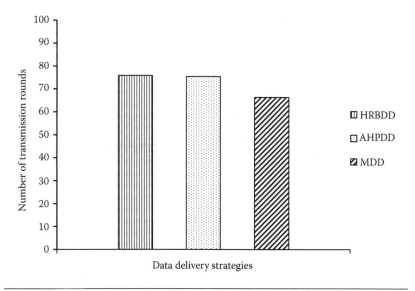

Figure 5.7   Comparison of the total number of transmission rounds.

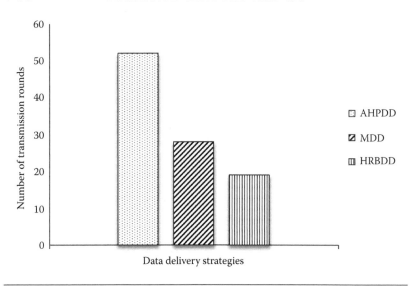

Figure 5.8 Comparison of the number of successful transmission rounds.

5.5.4 Use-Case Analysis of the Data Delivery Protocols Based on QoI Attribute Performance

To analyze the performance of the three data delivery techniques in terms of the QoI attributes, we hereby adopt a use case based on the simulations in Section 5.5.3. The remainder of this use case will refer to Figure 5.9 and Table 5.7.

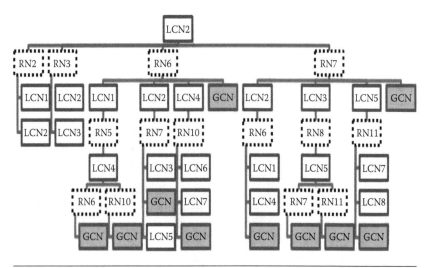

Figure 5.9 Tree-based illustration of a subset of paths from LCN2 to GCN through RNs 2, 3, 6, 7.

Table 5.7 Comparative Analysis of Data Delivery Paths in Terms of QoI Attributes

RN#	REMAINING BATTERY LEVELS			QoI ATTRIBUTES			EFFECTIVE QoI	CHOSEN NEXT HOP NODE			DATA DELIVERY PATH		
	AHPDD	HRBDD	MDD	LATENCY	RELIABILITY	THROUGHPUT	TYPE1 TRAFFIC	AHPDD	HRBDD	MDD	AHPDD	HRBDD	MDD
2	11	9	9	0.0219	0.7659	4.6606	0.1999	7	2	2,3,6,7	LCN2->RN7->GCN	LCN2->RN2->LCN1->RN5->LCN4->RN10->SINK	LCN2->RN6/7->SINK
3	9	9	7	0.0126	0.9958	8.1039	*0.3013*		Hop that offers best QoI=RN2				
6	7	7	5	0.0168	0.951	6.0936	0.2457						
7	3	5	3	0.0161	0.9619	6.3564	**0.2531**						
2	11	9	7	0.0126	0.9958	8.1023	0.2734	6	3	2,3,6,7	LCN2->RN6->GCN	LCN2->RN3->RN8->LCN5->RN7->SINK	LCN2->RN6/7->SINK
3	9	7	5	0.0246	0.5878	4.1472	0.1659		Hop that offers best QoI=RN6				
6	5	7	3	0.0117	0.9976	8.6991	***0.2873***						
7	3	3	1	0.0126	0.9958	8.1023	0.2734						
2	11	7	5	0.0203	0.8455	5.0411	*0.2752*	6	2	2,3,6	LCN2->RN6->GCN	LCN2->RN2->LCN1->RN6->SINK	LCN2->RN6->SINK
3	9	7	3	0.0196	0.8691	5.2057	0.2704		Hop that offers best QoI=RN2				
6	3	5	1	0.0168	0.9496	6.0932	**0.2478**						
7	3	3	1	0.012	0.9972	8.5106	0.2067						
2	11	7	3	0.0224	0.7383	4.5636	0.1963	7	3	2,3	LCN2->RN7->GCN	LCN2->RN3->RN7->SINK	LCN2->RN2->LCN1->RN5->LCN4->RN10->SINK
3	9	7	1	0.0117	0.9976	8.6991	*0.3178*		Hop that offers best QoI=RN3				
6	3	5	1	0.0224	0.7383	4.5636	0.1963						
7	1	1	1	0.0133	0.9926	7.6506	**0.2895**						

(Continued)

Table 5.7 (Continued) Comparative Analysis of Data Delivery Paths in Terms of QoI Attributes

RN#	REMAINING BATTERY LEVELS			QoI ATTRIBUTES			EFFECTIVE QoI	CHOSEN NEXT HOP NODE			DATA DELIVERY PATH		
	AHPDD	HRBDD	MDD	LATENCY	RELIABILITY	THROUGHPUT	TYPE1 TRAFFIC	AHPDD	HRBDD	MDD	AHPDD	HRBDD	MDD
2	11	5	1	0.0117	0.9976	8.6991	0.2758	6	2	2	LCN2->RN6->GCN	LCN2->RN2->LCN1->RN6->SINK	LCN2->RN2->LCN1->RN5->LCN4->RN10->SINK
3	9	7	1	0.0196	0.8691	5.2057	0.1957	Hop that offers best QoI=RN6					
6	3	3	1	0.0117	0.9976	8.6991	*0.2758*						
7	1	1	1										
2	11	5	1	0.0117	0.9976	8.6991	*0.2745*	6	3	—	LCN2->RN6->GCN	LCN2->RN3->LCN3->RN8->LCN5->RN11	LCN2 disconnected from Sink (failed transmission)
3	9	7	1	0.0133	0.9926	7.6506	0.2514	Hop that offers best QoI=RN2				RN11 is one-hop from Sink; assumed dead	
6	1	3	1	0.0161	0.963	6.3564	**0.2227**						
7	1	1	1	0.0133	0.9926	7.6506	0.2514						

LCN2 is identified as the source node that has data to be delivered to the GCN, in response to periodic requests (Traffic Type 1) during each transmission round. The one-hop neighbor RNs of LCN2 are RN2, RN3, RN6, and RN7, and have battery levels of 11, 9, 7, and 5 units, respectively, at the start of the simulation instant. Values of the QoI attributes are recorded for each of the one-hop RNs. AHP analysis is performed to identify the best forward hop RN for AHPDD as indicated in bold under the column titled "Effective QoI." The theoretical best next hop RN for the other two protocols is found using AHP analysis (highlighted in green), to compare the QoI performance of the actual next hop node chosen by the other two protocols. Comparing the QoI performance of the chosen next hop node, we make the following observations: AHPDD always chooses the best QoI providing node between RN6 and RN7, as long as they are available. Although RN2 or RN3 might provide better QoI values for the next hop in some cases, choosing the forward hop RNs reduces the number of hops to reach the GCN. This leads to lesser energy consumption in the network on the whole, and also reduces the cumulative latency along the data delivery path to the GCN. However, this also means that once the forward hop RNs die out, AHPDD has to make use of longer data delivery paths to the GCN. But again, the QoI attributes are still considered in choosing the best among the available next hop nodes. MDD on the other hand, is always able to deliver data through at least one next hop node that provides the best effective QoI for each traffic type, even though it doesn't have a mechanism to identify the best next hop node.

It is also able to find the shortest route to the sink because of the multipath approach at each next hop node. However, this performance comes at the cost of a higher overall energy consumption in the network. This can be seen from Table 5.7, where all the one-hop nodes run out of energy before the other two techniques. Comparing with the observations made from Figures 5.6 through 5.8, we see that although MDD lasts for lesser number of transmission rounds, not only does it provides a lower failure rate, it also performs well in terms of identifying at least one next hop node that provides the best QoI performance. As for HRBDD, what stands out from Table 5.7 is the increased number of hops in delivering data to the sink, causing an overall increase in energy consumption in the network.

This is because HRBDD is always trying to find a node with higher remaining energy at each next hop, irrespective of its QoI performance. Although the chosen next hop node sometimes provides the best QoI, HRBDD's performance with respect to QoI attributes is not consistently good. Over a period of time, this leads to death of more intermediate nodes, causing a higher failure rate as indicated by Figure 5.7, as the sink cannot be reached along a chosen path. This leads to lesser number of successful transmissions to the sink, even though the network might be able to run for a little longer than the multipath routing technique, as shown by Figure 5.8 and Figure 5.6, respectively. Thus, HRBDD performs relatively poorly among the three data delivery strategies, both in terms of delivering data with user-desired QoI attributes, and in terms of the number of successful transmission rounds.

5.6 Conclusions

In this chapter, we proposed a framework for cognitive information-centric sensor networks (ICSN) that can be used to implement information-centric data delivery using elements of cognition, that is, knowledge representation and inference to advance data-centric sensor networks to cognitive information-centric sensor networks. These ICSNs are able to handle heterogeneous traffic flows in the network generated as a result of requests coming from multiple clients in SOM applications while considering the QoI attribute priorities for each traffic flow. From the simulations we were able to identify the number of sensor nodes that should be simultaneously scheduled while gathering data to ensure good quality data from the sensor nodes. Optimally choosing the number of simultaneously transmitting sensor nodes improves the average throughput by about 85 percent, reliability by about 90 percent, and reduces the latency by about 18 percent for a given value of offered load (1000 bits). The simulation-generated values were used in the next set of simulations that implemented AHP analysis to decide the best next hop node that should be used for data delivery to the GCN. It was found that the network lasted for significantly more number of transmission rounds, and performed well in responding to varying traffic types and changing network topology when it implemented cognitive routing decisions, when compared with traditional decision techniques. In our future work, we will enhance the learning strategy, and implement cache replacement at LCNs to further exploit the cognitive node's capabilities to improve network performance and prolong the network lifetime, while meeting the end user's requirements.

References

1. Al-Fagih, A., Al-Turjman, F., Alsalih, W., and Hassanein, H., "A Priced Public Sensing Framework for Heterogeneous IoT Architectures," *IEEE Transactions on Emerging Topics in Computing*, vol. 1, no. 1, pp. 135–147, Oct. 2013.

2. Ahlgren, B., Dannewitz, C., Imbrenda, C., Kutscher, D., and Ohlman, B., "A Survey of Information-Centric Networking," *Communications Magazine*, IEEE, vol. 50, no. 7, pp. 26, 36, July 2012. DOI: 10.1109 /MCOM.2012.6231276.

3. Al-Turjman, F., and Hassanein, H., "Enhanced Data Delivery Framework for Dynamic Information-Centric Networks (ICNs)," in *Proc. of the IEEE Local Computer Networks (LCN)*, Sydney, Australia, 2013, pp. 831–838.

4. Ahmed, K., and Gregory, M.A., "Techniques and Challenges of Data Centric Storage Scheme in Wireless Sensor Network." *J. Sens. Actuator Netw*, vol. 1, no. 1, pp. 59–85, 2012. DOI: 10.3390/jsan1010059.

5. Stojmenovic, I., Ed., *Handbook of Sensor Networks: Algorithms and Architectures*, John Wiley & Sons, Inc., Hoboken, NJ, 2005.

6. Krishnamachari, B., Estrin, D., and Wicker, S., "Modelling Data-Centric Routing in Wireless Sensor Networks," IEEE Infocom, vol. 2, pp. 39–44, June 2002.

7. Intanagonwiwat, C., Govindan, R., Estrin, D., Heidemann, J., and Silva, F., "Directed Diffusion for Wireless Sensor Networking," *Networking, IEEE/ACM Transactions on*, vol. 11, no. 1, pp. 2–16, 2003.

8. Ratnasamy, S., Karp, B., Shenker, S., Estrin, D., Govindan, R., Yin, L., and Yu, F., "Data-Centric Storage in Sensornets with GHT, a Geographic Hash Table," *Mobile Networks and Applications*, vol. 8, no. 4, pp. 427–442, 2003.

9. "ZigBee Specification," Jan. 2008. Available at http://www.zigbee.org ZigBee Document 053474r17.

10. Vijay, G., Ben Ali Bdira, E., and Ibnkahla, M., "Cognition in Wireless Sensor Networks: A Perspective," *Sensors Journal, IEEE*, vol. 11, no. 3, pp. 582–592, 2011.

11. Shenai, K., and Mukhopadhyay, S., "Cognitive Sensor Networks," *Proc. IEEE 26th Int. Conf. Microelectronics* (MIEL), pp. 315–320, 2008.

12. Reznik, L., and Von Pless, G., "Neural Networks for Cognitive Sensor Networks," *Proc. IEEE Int. Joint Conf. Neural Network*, IJCNN, pp. 1235–1241, 2008.

13. Sachidananda, V., Khelil, A., and Suri, N., "Quality of Information in Wireless Sensor Networks: A Survey," 15th Int'l Conf. on Information Quality (ICIQ 2010). Little Rock, AK, pp. 193–207, 2010.

14. Bisdikian, C., Kaplan, L. M., and Srivastava, M. B., "On the Quality of Information in Sensor Networks," *ACM Trans. Sensor Netw*, vol. 9, no. 4, article 48, July 2013. DOI: http://dx.doi.org/10.1145/2489253.2489265.

15. Cheriton, D., and Gritter, M., "TRIAD: A New Next-Generation Internet Architecture," Jan. 2000.

16. Al-Turjman, F., Alfagih, A., and Hassanein, H., "A Value-Based Cache Replacement Approach for Information-Centric Networks," in *Proc. of the IEEE Local Computer Networks (LCN)*, Sydney, Australia, 2013, pp. 902–909.

17. Ming, Z., Xu, M., and Wang, D., "Age-Based Cooperative Caching in Information-Centric Networks," *Computer Communications Workshops (INFOCOM WKSHPS), 2012 IEEE Conference on*, pp. 268–273, March 2012. DOI: 10.1109/INFCOMW.2012.6193504.

18. Heinzelman, W. B., Chandrakasan, A. P., and Balakrishnan, H., "An Application-Specific Protocol Architecture for Wireless Microsensor Networks," *Wireless Communications, IEEE Transactions on*, vol. 1, no. 4, pp. 660–670, Oct 2002. DOI: 10.1109/TWC.2002.804190.

19. Alsbou, T. A. A., Hammoudeh, M., Bandar, Z., and Nisbet, A., "An Overview and Classification of Approaches to Information Extraction in Wireless Sensor Networks," in *Proc. of the 5th Intl. Conference on Sensor Technologies and Applications (SENSORCOMM '11)*, Nice, Saint Laurent du Var, France, IARIA, 2011.

20. Akbas, M. I., and Turgut, D., "Lightweight Routing with Dynamic Interests in Wireless Sensor and Actor Networks," *Elsevier Ad Hoc Networks*, vol. 11, no. 8, pp. 2313–2328, November 2013.

21. Clark, D. D., Partrige, C., Ramming, J. C., and Wroclawski, J. T., "A Knowledge Plane for the Internet," *Proc. SIGCOMM 2003*, pp. 3–10, 2003.

22. Thomas, R. W., Friend, D. H., DaSilva, L. A., and MacKenzie, A. B., "Cognitive Networks: Adaptation and Learning to Achieve End-to-End Performance Objectives," *IEEE Commun. Mag.*, vol. 44, no. 12, pp. 51–57, 2006.

23. Boyd, J., "A Discourse on Winning and Losing: Patterns of Conflict," 1986.

24. Vijay, G., and Ibnkahla, M., "CCAWSN: A Cognitive Communication Architecture for Wireless Sensor Networks," in *Proc. 26th Biennial Symposium on Communications*, QBSC 2012, pp. 132–137.

25. Al-Turjman, F., Hassanein H., and Ibnkahla, M., "Towards Prolonged Lifetime for Deployed WSNs in Outdoor Environment Monitoring," *Elsevier Ad Hoc Networks Journal*, vol. 24, no. A, pp. 172–185, Jan. 2015. DOI: 10.1016/j.adhoc.2014.08.017.

26. Al-Turjman, F., Hassanein H., and Ibnkahla, M., "Quantifying Connectivity in Wireless Sensor Networks with Grid-Based Deployments," *Elsevier: Journal of Network & Computer Applications*, vol. 36, no. 1, pp. 368–377, Jan, 2013.

27. Al-Turjman, F., Hassanein H., and Ibnkahla, M., "Efficient Deployment of Wireless Sensor Networks Targeting Environment Monitoring Applications," *Elsevier: Computer Communications Journal*, vol. 36, no. 2, pp. 135–148, Jan. 2013.

28. Chen, D., and Varshney, P. K., "QoS Support in Wireless Sensor Networks: A Survey," *Proc. Intl. Conf. on Wireless Networks* (ICWN), 2004.

29. Park, P., Di Marco, P., Soldati, P., Fischione, C., and Johansson, K. H., "A Generalized Markov Chain Model for Effective Analysis of Slotted IEEE 802.15.4," *Mobile Adhoc and Sensor Systems, 2009. MASS '09. IEEE 6th International Conference on*, vol. 130, no. 139, pp. 12–15, Oct. 2009. DOI: 10.1109/MOBHOC.2009.5337007.

30. Singh, G., and Al-Turjman, F., "Learning Data Delivery Paths in QoI-Aware Information-Centric Sensor Networks," *IEEE Internet of Things Journal*, vol. 3, no. 4, pp. 572–580, 2016.

31. ITU-T Series Y recommendation: ITU-T Y.2221, "Requirements for Support of Ubiquitous Sensor Network Applications and Services in the NGN Environment," Jan. 2010.

32. Haykin, S., "Cognitive Radio: Brain-Empowered Wireless Communications," *IEEE J Sel Area Comm.*, no. 23, pp. 201–220, 2005.

33. Mitola, J., and Maguire, G. Q., "Cognitive Radio: Making Software Radios More Personal," *IEEE Personal Communications*, vol. 6, no. 4, pp. 13–18, 1999.

34. Friend, D. H., Thomas, R. W., MacKenzie, A. B., and DaSilva, L. A., "Distributed Learning and Reasoning in Cognitive Networks: Methods and Design Decisions," in *Cognitive Networks: Towards Self-Aware Networks* (Q. H. Mahmoud, ed.), pp. 223–246, John Wiley & Sons, Hoboken, NJ, 2007.

35. Singh, G., Abu-Elkheir, M., Al-Turjman, F., and Taha, A., "Towards Prolonged Lifetime for Large-scale Information-Centric Sensor Networks," in *Proc. of the IEEE Queen's Biennial Symposium on Communications (QBSC)*, Kingston, Ontario, Canada, 2014, pp. 87–91.

36. Singh, G., and Al-Turjman, F., "Cognitive Routing for Information-Centric Sensor Networks in Smart Cities," in *Proc. of the International Wireless Communications and Mobile Computing Conference (IWCMC)*, Nicosia, Cyprus, 2014, pp. 1124–1129.

37. Analytic Hierarchy Process, Wikipedia, available at http://en.wikipedia.org/wiki/Analytic_hierarchy_process.

38. Steels, L., "Frame-Based Knowledge Representation," Working paper 170, MIT AI Laboratory, Cambridge, MA, 1978.

39. Stengel, R., "Lecture Slides on 'Knowledge Representation,'" available at http://www.princeton.edu/~stengel/MAE345Lectures.html.

40. Dimokas, N., Katsaros, D., Tassiulas, L., and Manolopoulos, Y., "High Performance, Low Complexity Cooperative Caching for Wireless Sensor Networks," *J. Wireless Networks*, vol. 17, no. 3, pp. 717–737, April 2011.

41. Zayani, M.-H., and Gauthier, V., "Usage of IEEE 802.15.4 MAC–PHY Model," available at http://www-public.it-sudparis.eu/~gauthier/Tools/802_15_4_MAC_PHY_Usage.pdf.
42. Zayani, M.-H., Gauthier, V., and Zeghlache, D., "A Joint Model for IEEE 802.15.4 Physical and Medium Access Control Layers," in *Proc. of IEEE The 7th International Wireless Communications and Mobile Computing Conference* (IWCMC 2011), 2011.

6

Cognitive Routing Protocol for Disaster-Inspired WSNs in the Internet of Things*

6.1 Introduction

Sensing technology has played a significant role in detection and containment of disasters in numerous disciplines. The most notable among these applications are those functioning in harsh environments, such as pollution and flood detection, forestry fire prevention and earthquake monitoring applications [1,2]. For instance, a network of sensors can be used to monitor the motion of toxic gasses over vast areas [3]. In [4], redwood trees at risk have been monitored via a wireless sensor network. Furthermore, the existence of a disinfectant that works better and more efficiently than conventional traditional ones by providing long lasting antiviral effect against major viruses has proved the importance of sensing technology in disaster management [5]. The aforementioned examples are just few of the many areas where sensing technology has made massive improvements. However, this technology is still suffering extreme limitations in terms of energy and connectivity while collaborating in wireless network–based systems.

Connectivity and links between sensing-devices (nodes) distributed to monitor a specific phenomenon have led to the idea of WSNs followed by the Internet of things (IoT) proposal. Integrating sensing-devices with other heterogeneous network systems such as WiFi,

---

* This article was originally published in *Future Generation Computer Systems*. F. Al-Turjman, Cognitive routing protocol for disaster-inspired WSNs in the Internet of Things, vol. 1, no. 1, 2017. Reprinted with permission.

LiFi, LTE, and so on can significantly expand the array of services that can be provided to public users as well as decision makers in critical safety applications. However, several design aspects, such as the limited energy constraints, short communication range between geolocated objects, and low processing power, need to be incorporated into the routing protocol in order to realize the IoT paradigm. There have been several attempts in the literature to propose lightweight solutions for the IoT paradigm in order to save energy [6–8]. However, these solutions are still under investigation. WSNs in IoT consume energy in almost all processes [9]. They consume energy while making data transmission, data sensing, and data processing. A few attempts toward achieving energy efficiency in such networks via wireless multi-hop networking have been proposed, for example, [10–12]. However, such schemes either assume static network topology, which render these schemes impractical for real-life network implementations, given that IoT-based networks exhibit random topology due to the mobility of nodes, or are restricted to two-hop from source to the sink. Due to the fact that IoT-networks' sensors are usually limited in their processing power, communication range and energy capabilities, design and implementation of routing algorithms are considered a nontrivial task.

Moreover, accommodating various levels of harshness in surroundings in terms of temperature, dust, humidity, and so on dictates a resilience requirement to an expanded set of failure possibilities that includes partial or complete failure of the IoT sensor nodes, and reduced levels of activity or accuracy undergone as batteries deplete, which forms serious threats for losing critical in-network data before being utilized. This requirement stresses the integrity of cognition elements in IoT routing protocols. Moreover, an IoT sensor network needs to sustain different levels of mobility on disaster situations [13]. Since the IoT connected things rely on each other to gather and process data, mobility may be temporarily or permanently detrimental to the network operation by breaking some functional communication links that affects the in-network data retrieval. Hence, nodes and links are prone to several risks, leading to high probabilities of failures and several nodes in the network may become disconnected/failed. As a reason of that, we are characterizing such circumstances by the probability of node failure (PNF). Consequently, for a successful and reliable operation of the IoT paradigm in disaster-inspired scenarios, a cognitive energy-efficient routing approach shall be applied.

Recently, there have been significant attempts in the direction of building a cognitive sensor network, where researchers have made use of artificial neural networks, genetic algorithms, game theory and even software agents to implement distributed and intelligent decision making in sensor networks [2,14,15]. However, there is no single framework that can be used to implement cognition in sensor networks supporting the IoT paradigm in a way that is domain- and application-independent.

In this chapter, we propose a cognitive data delivery approach that addresses the challenges of data delivery in IoT networks comprised of energy-constrained IoT sensors. Two key elements in cognition are utilized in our approach in order to implement cognition, *reasoning* and *learning* elements. Reasoning is used to differentiate between the attributes of a given traffic flow, and choosing the next hop along the data delivery path to the destination. While reasoning realizes short-term objectives and makes decisions based on the current network status, learning assists in achieving long-term goals of the network, such as improving its lifetime. The feedback received from the previous history of the exchanged messages aids the learning process, and leads to proactive actions. This model will help us to specify which path to follow in order to determine the optimal usage of the available resources for a wide range of IoT applications in safety and security scenarios. Where the proposed approach is energy-efficient and designed to optimize the current network status for guaranteed quality of service (QoS) via machine learning [14]. It provides efficient and self-healing data delivery while choosing and selecting reliable communication links [15]. Moreover, the proposed approach caters for the grid-based distribution of the employed IoT sensors on the monitored object to efficiently and effectively cope with the dynamicity of IoT-network topology [16].

The reset of the paper is organized as follows. Section 6.2 reviews previous related work in the literature. Section 6.3 clarifies our system models. Section 6.4 describes our proposed routing approach for disaster-inspired IoT paradigms. In Section 6.5, performance evaluation results for the proposed approach in comparison to other related approaches are detailed. Finally, Section 6.6 provides concluding remarks.

6.2 Related Work

Connectivity in IoT relies on finding reliable routes from IoT "things" to the Internet gateway. Utilizing duty-cycling, a routing algorithm can be designed to significantly balance the network load and optimize the energy consumption, especially in energy-constrained IoT networks. In mission-critical IoT networks, it is also important when designing the routing algorithms to facilitate prioritization between the different traffic types. Another critical problem to overcome is the uneven energy consumption across the network where elements near the gateway would deplete their energy faster than those that are far away. And hence, feedback from the MAC and physical layers, in addition to information about the residual energy and the current load of the distributed nodes, can be utilized to identify and avoid unreliable links in order to effectively prolong the network lifetime and increase the network throughput. There are considerable advantages in coupling an IoT routing protocol with the underlying MAC layer protocol through a cross-layer design. Reducing the ratio of lost packets during channel impairments is an important reliability objective, as well [17]. Where energy saving via cognitive radio can be applied at the MAC layer of the network stack to achieve this target [18,19]. The MAC layer is usually expected to adapt the number of retransmissions depending on channel quality. Current MAC protocols typically limit the number of back-offs and retransmissions. However, the unique characteristics of the short-range communications and the specific challenges of IoT mainly energy and processing constraints in addition to the random network topology, prevent the direct implementation of traditional WSNs' routing schemes without modification. In the following, we discuss traditional WSNs' protocols with preferable features to be considered in an IoT routing protocol.

Multipath versus single-path routing: Recent studies show that multipath routing protocols for sensor networks are better than single-path routing protocols in terms of QoS. In fact, multipath routing protocols provide lower probability of packet loss while utilizing redundant paths towards the sink [20]. The Reliable Information Forwarding using multiple paths (ReInForM) protocol as described in [21] employs a probabilistic flooding to deliver information awareness packets with desired priority levels of reliability at proportional costs for sensor networks.

This routing mechanism is based on local knowledge of network conditions, such as channel error, hops-to-sink counting, and connectivity degree. Unfortunately, this protocol is not designed specifically for real-time traffic; therefore, it does not consider delay deadlines of packets when selecting the multiple paths. A chosen path might not be able to meet the delay requirements, yet it will be used to propagate duplicates potentially consuming valuable energy and unduly occupying useful channel bandwidth without improving the system performance. The work in [22] presents a data flooding dissemination scheme. It assumes a virtual-grid network architecture where sensor nodes are distributed densely at the vertices of the grid. Utilizing the uniform nodes' patterns and lattice algebra, the scheme dismisses node-addressing requirements and employs a simple flooding scheme for data dissemination. While the proposed scheme simplifies the communication model, it overlooks the cost of real-time signal processing. Additionally, it assumes a fixed structure and a static node deployment. Nevertheless, IoT-nodes can move around us for certain health or traffic applications [23], and therefore, may need to be associated with different neighbors and may not always follow a fixed structure. Authors of [24] propose a sound routing scheme for energy harvesting in IoT-networks. The routing scheme assumes a hierarchical cluster-based architecture. Packet transmission from the source to the cluster head via single or multi-hop fashion. Still, however, the challenge of limited energy budget at the sensing-node is not considered effectively.

Geo-based routing protocols: These protocols utilize the node position information in order to achieve more efficient routing techniques. For example, the geographic adaptive fidelity (GAF) protocol optimizes the performance of the sensor network by determining the redundant nodes via precisely identify their geographical positions [25]. Where these redundant nodes are considered equivalent and useful in terms of relaying/forwarding packets in the network. Another routing protocol, namely the GEAR (Geographic and Energy-Aware Routing), aims at improving the energy efficiency by forwarding queries to specifically determined regions [26]. In this routing protocol, sensors need to have localization hardware such as a GPS unit or a localization system which can dramatically increase the network cost. Meanwhile, the work presented in [27] proposes a geographic routing protocol where the IoT-nodes are assumed to include two types of anchor nodes, which have higher communication and processing capabilities than user nodes. User nodes are required to localize their position with reference to these anchor nodes. However, the proposed scheme is topology-dependent and assumes a fixed topology, which may not be applicable for IoT networks. Additionally, the scheme requires addressing for all nodes, which forms a significant challenge in IoT networks with large-scale applications. Moreover, the localization techniques used can be inaccurate and lead to dramatic degradation in energy consumption.

Shortest Path Routing: In the nearest neighbor algorithm (NNA), when a packet is transmitted from a node to another, it follows the shortest path based on the available common control channels. NNA assumes that if a packet always follows the shortest path, it will use it until it reaches destination node. In short, this algorithm uses a four-direction transmission (left, right, up, and down) only so it actually does not consider the shortest path, it considers the shortest neighbor relay node in order to send data. As a result of this, hop count unnecessarily increases and therefore energy consumption is negatively affected. Meanwhile, in the shortest path algorithm (SPA), when a packet is transmitted from a node, the algorithm calculates the shortest path from recent node to destination instead of node-to-node, and the packet follows this path until it reaches to destination. SPA, uses eight directions (up, upper-left, upper-right, down, down-left, down-right, right, and left) and considers the shortest path to the destination rather than the shortest neighbor of the relay node. Thus, in SPA hop count decreases and energy consumption of the nodes decreases in comparison with NNA. Nevertheless, SPA is the simplest routing protocol that takes into consideration the path length as a unique design factor affecting the network energy. In practice, this assumption is not accurate due to several other design factors, such as the communication link condition and reliability.

In this research, we propose a cognitive energy-efficient algorithm (CEEA) as a routing protocol. CEEA assumes a multitier IoT-network, and cluster/tier-wide synchronization. It is a topology-independent protocol which copes with the randomness nature in IoT-networks. CEEA determines the path from the routing node (RN) to the destination node in view of each node's remaining energy. The remaining energy of neighbors of recent RNs is controlled each time before a packet is sent from the RN. If one of the neighbor RN's energy is below half of its initial value, a new path will be determined for the packet to follow. In addition, when all neighbors' remaining energy is below half of the initial energy, the system uses the same strategy. As a result of that, even if hop count increases in comparison with SPA, energy efficiency is improved for RNs and so is the network lifetime.

6.3 System Model

The main objective of IoT in smart environments is to monitor physical or chemical changes and pass the information to a data center for processing [28]. IoT nodes may have varying sensing capabilities. Due to energy constraints, nodes do not communicate with each other but rather pass the sensed data to routing nodes (RN). RNs take collected data to a gateway (GCN) that is usually connected to the Internet for remote collection/processing. This communication type continues till the IoT network death. IoT networks have to overcome several challenges. Energy consumption for communication is the most significant one. Energy-efficient routing protocols can significantly prolong the IoT-network lifetime. In this section we list the assumed system model for the proposed CEEA routing protocol.

6.3.1 Network Architecture

The typical communication range in IoT is expected to be between 1 centimeter and 150 meters [9]. This means that the transmission range is still limited, making multi-hop routing particularly important for IoT networks. Furthermore, when IoT nodes are mobile, the direction of a communication route is not deterministic and is dependent on the drift velocity of sensory machines, which may lead to communication delay. This necessitates efficient schemes for multi-hop path creation and management. IoT networks can be divided into three categories: in-object, on-object, and off-objects IoT. An overview of the structure of IoT network under such circumstances can be summarized as

- IoT Sensor Nodes (SNs): These are assumed to be small and simple IoT sensor devices. Due to their limited energy, limited memory, and reduced communication capabilities, they can only perform simple computation task and can transmit over very short distances. The nodes could be composed of sensor and communication units.

- Relay Nodes (RNs): These are the relay (routing) devices with slightly larger computational resources than SNs and can aggregate information from a limited number of SNs and also can control the behavior of SNs by sending simple instructions (such as on/off, sleep, read value, etc.). These added capabilities would increase their size; thus, their deployment would be more invasive.
- Cognitive Relay Nodes (CRNs): They are used to aggregate the information forwarded by RNs and send the information to other CRN devices. At the same time, they can send the information from short-range–scale to large-scale. In this chapter we identify these nodes as cognitive nodes (CRNs).
- Gateway (GCN): It enables to control or monitor the entire IoT system remotely over the Internet.

It's worth pointing out here that IEEE 802.15.4 protocol is considered at the CRN to specify a sublayer for medium access control (MAC) and a physical layer (PHY) for low-rate wireless private area networks (LR-WPAN) because of some desired features, such as low power consumption, low data rate, low cost, and high message throughput [9]. Thus, the IEEE 802.15.4–based CSMA access method can be considered at the MAC layer. This inherently reflects the communication channel reliability. Based on [29], this channel reliability can be characterized by a reliability design factor as follows:

$$C_R = ((1 - P_{blocking}) * (1 - P_{c-fail}) * (1 - P_{p-discard})) \qquad (6.1)$$

Where $P_{blocking}$ represents the blocking probability due to a buffer-full condition; P_{c-fail} is the common channel access failure probability due to channel condition (i.e., SNR) and $P_{p-discard}$ is the probability that a packet is discarded on reaching the maximum number of retries limit. This reliability factor is responsible to make a decision when equivalent energy levels at RNs are faced. And it reflects the probability that a frame is not blocked, lost due to common channel access failure, or discarded as a result of reaching the maximum number of retries limit.

6.3.2 Lifetime of IoT Network

IoT Network Lifetime is defined as the time or number of transmission rounds beyond which the network can no longer deliver useful information to the outside end user. This is reflected by the network's inability to find a data delivery path with satisfactory values for quality of information (QoI) attributes such as delay, reliability, and throughput, as determined by the end user [24]. This definition not only provides information to satisfying the application requirements, but also considers the status of the network and sensing resources in defining the network lifetime. It also justifies the fact that if the network does not have the necessary resources to send packets, it cannot satisfy the end user, and so it should be considered as a dead IoT network. The IoT network lifetime can therefore be evaluated in three ways:

1. Lifetime Based on Number of Alive SNs.

   Several variants do exist with this model. The simple model identifies the time until the death of the first SN in the network as the lifetime of the network. Another variant evaluates lifetime until the death of k out of n SNs in the network, where $k < n$. The lifetime is the range between the death of k nodes from n nodes in noncritical ones [30].

2. Lifetime Based on SN Coverage.

   This model defines the lifetime of the network in terms of the coverage of region of interest. If it is used to ensure that all points inside a region of interest are covered, it is denoted by volume coverage. When an identified number of target points are to be covered, it is denoted as target coverage.

3. Lifetime Based on Coverage and Alive SNs.

   This type of metrics is mostly found in ad-hoc IoT networks. In this option, *lifetime* is defined as the period during which most of the nodes are connected with each other. Because in IoT each node has to communicate with a gateway node, this metric cannot be used as-is. Another issue with this metric is that the lifetime is based on the total number of packets transmitted to the gateway. Nevertheless, in most of the related works this metric become useless [31].

6.3.3 Energy Conservation and Dead Node Issue

Energy conservation is one of the most important issues in IoT design. SNs are restricted in carrying out the network layer functions, their main task is to flood the data to their one hop routers. Hence, the multi-hop forwarding between source and gateway is normally performed by RNs which have relatively higher capabilities than SNs. Thus, we define the energy consumed at a RN by $E_{RN} = C\,(T * (E_{TX}) + R * (E_{RX}))$. Most of the energy consumption at the RN is due to data communication, indicated by E_{TX} for energy consumed during transmission and E_{RX} for energy consumed during data reception. C represents the cost function of the energy consumed T represents the number of transmitted packets and R represents number of received packets. As discussed above, CRN main function is data aggregation and routing of traffic received from the RNs via cognition elements. The capabilities of the CRN are higher than those of RN, hence our assumption of the cognitive decision process to be performed by the CRNs, which is expected to consume additional energy compared to regular RNs [32]. Additional energy consumption is divided into two parts: One is protocol overhead incurring during cognitive data delivery due to feedback from the IoT network during the learning process and the exchange of values of QoI attributes such as delay, reliability, and throughput while making routing decisions; the other one is the increased transmit power for increasing the communication range of CRNs. Accordingly, $E_{CRN} = C\,(T * (E_{TX}) + R * (E_{RX})) + C(Ag * (E_{ag})) + C\,(P * (E_{cog} - E_{pro}))$, where T, R, Ag, and P represents the total number of packets that are transmitted, received, aggregated, and processed by the cognitive elements, respectively, in each transmission round. $(T * (E_{TX}) + R * (E_{RX}))$ is the energy cost incurred during data transmission and reception, $C(Ag * (E_{ag}))$ represents the energy cost incurred during data aggregation and $C(P * (E_{cog} - E_{pro}))$ indicates the energy cost due to protocol and processing overhead during the cognitive processes. Consequently, we can assume

$$E_{CRN} \geq E_{RN} + (Ag * (E_{ag}) + C(E_{cog} - E_{pro})) \qquad (6.2)$$

If the RN and CRNs use the same transmit power, the equality sign becomes positive in Equation 6.2. In order to ensure that the energy cost of CRNs does not offset the advantages it offers in terms of adapting to traffic flow dynamics and network topology alterations, the cost can be optimized by maximizing the number of RNs and minimizing the number of CRNs in the deployment [24].

In this work, we refer to one-hop neighbors' communication as the first tier of nodes. Since no other node can reach the monitoring station directly, traffic from every other node will have to be forwarded, in the last hop, by one of these first tier nodes. Similarly, the two-hop neighbors of the monitoring station will forward data for all nodes except the one-hop neighbors and themselves, and so on. If the spatial distribution of nodes is assumed to be uniform, then the traffic load is equally distributed. Each first tier node will forward hardly the same amount of traffic, and all first tier nodes will die at times very close to each other after the network is first put into operation. Once all of the first tier nodes are dead, no other node will be able to send data to the gateway node, and the lifetime of the network will be over. Increasing the number of nodes in the network accentuates this effect, since there is more traffic to forward and the first tier of nodes has a smaller share of the total energy budget. In general, the network death in IoT can be associated with several cutoff criteria such as the first node death, the percentage of dead nodes, or the number of dead nodes rising above a level where the routing to the sink node is no longer possible [33]. Nevertheless, as we are experimenting with the clustering based protocols, in which the energy is evenly distributed throughout the mobile IoT network, we consider the first scenario for the definition of the network lifetime. Because, when the first node dies, the number of dead nodes increases in the later rounds, and within 5–10 rounds the whole network becomes nonoperational. According to preliminary results, non-position–based routing protocols outperform geo-based protocols in terms of network lifetime. The primary reason for this behavior is that location-based protocols consume energy in terms of localization services. Moreover, the number of control messages plays a vital role in the network lifetime.

6.3.4 Communication Model

Radio interference, antenna shape and orientation, distance and environmental factors may vary during the network lifetime and affect link quality between the sensor nodes [16]. Despite the fact that the

locations of sensor nodes are fixed as well as every node is config-
ured with the same transmission range, environmental variations
result in asymmetric links between nodes [34]. Therefore, these routing approaches shall estimate link quality to find the optimal path.
Considering that the communication is at varying-range scale, study
of the communication in a very short range is essential. And hence,
we consider the proposed path loss formula in [33] at short-range
communication, which has two parts: The absorption path loss and
the spread path loss. Meanwhile, energy-aware frameworks depend
heavily on two main principles in their communication design. First
principle is the number of hops without delay constraint. Second, is
the number of hops with the delay constraints.

6.3.4.1 Number of Hops without Delay Constraints   If there is no delay
constraint on the system, the highest achievable transmission rate is given
by equation 18 in 0. The bandwidth is divided into i sub-bands the i-th
sub-band is centered around frequency $f_i, i = 1, 2,...$ and it has width Δf.
If the sub-band width is small enough, the channel appears as frequency-
nonselective and the noise p.s.d. can be considered locally flat. The resulting capacity in bits/s is then given by where d is the total path length, S
is the transmitted signal p.s.d., A is the channel path loss, and N_0 is the
noise p.s.d. The end-to-end capacity of N hops path is given by [35]:

$$C_{e2e} = C_I(1 - F_{AVG})^N \tag{6.3}$$

where C_I is the channel capacity contributed by the first hop, N is the
hop count determined by the forwarding scheme and F_{AVG} is the average
capacity loss factor per hop. The value of F_{AVG} is calculated as follows:

$$F_{AVG} = F\left(\frac{d_0 - d_{AVG}}{d_0}\right) \tag{6.4}$$

where F is the capacity loss factor and d_0 is a constant that denotes the
reference distance from source-to-sink 0.

6.3.4.2 Number of Hops with Delay Constraints   The predictions about
the preferred number of hops made in the previous section were based
on the assumption that the block lengths used by channel codes can

be randomly large. In many applications there is a strict limit on the tolerable end-to-end delay. There are several factors of delay in short-range communication systems. In the following we list these factors:

- *Waiting* for the data source to emit enough bits to form a block of a desired length (for channel coding).
- *Processing* delay caused by encoding/decoding the information bits for transmission.
- *Transmission* and *reception* of the whole encoded message.

If the communication system involves multiple hops, the latter three elements are repeated several times, increasing overall delay. To compensate for this, shorter block lengths must be used at a cost of reduced error-correcting capabilities at each link [24].

6.4 Cognitive Energy-Efficient Approach (CEEA)

In this section we propose a novel energy aware data delivery approach for the energy-constrained IoT. Let's assume that a randomly selected sensor n by GCN, depending on the harvested energy, is to be used for data retrieval. The random number of relay nodes within the communication range of the sensor n can be characterized by a spatial Poisson process X. Let the sensor n be at point $z \in \mathbb{R}^2$ and define $l(z, X)$ as the shortest distance from the sensor location z to the nearest point of X such that $l(z, X) \leq r$, and only common control channels are considered. Since X is a spatial Poisson process, then $l(z, X) \leq r$, if and only if $RN(d(z, r)) > 0$, where $d(z, r)$ is a disc of radius r centered at z. Conclusively, the probability of having at least one relay neighbor within the transmission range of the sensor n is given as follows.

$$P(l(z, X) \leq r) = 1 - exp(\beta_{harv} A_d(d(z, r)))$$ (6.5)

where A_d is the area of the disk $d(z, r)$, and where β_{harv} denotes the rate of harvested energy of a senor. Note that $exp(\beta_{harv} A_d(d(z, r)))$ denote the probability that no relay node is within the transmission range of the sensor n; that is, the network lifetime of the neighborhood of sensor n is expired. When the lifetime of the neighborhood nodes is expired, the IoT-network is assumed dead. Thus, assuming $f(n_j)$ is the cost function of transmitting from RN_j to GCN, $g(n)$ is the energy of neighboring RNs, $h(n)$ is the minimum distance from a neighbor RN_j to GCN, $i(n)$ initial energy of the neighboring RN. Our approach relies on a *cognition process* that has three main criteria in data routing: (1) Evaluation criteria, $f(n_j)$ = Cost(Neighbor RN to GCN) and $h(n_j)$ = $min(f(n_j))$, this is guaranteed by lines 11 to 18 in Algorithm 6.1a; (2) selection criteria, $g\ (h(n_j)) > i\ (h(n_j)) * 50\%$, this section is found between lines 19 and 21; and (3) termination Criteria; all one-hop RNs are dead or $P(l(z, X) \le r) = 0$.

Algorithm 6.1a Pseudo-Code of the CEEA Algorithm

1. **Function: Cognition process in CEEA**
2. **Input**
3.   Source RN
4. **Output**
5.   RN index chosen by CEEA to deliver data towards GCN.
6. **Begin**
7. **Initialize**
8.   Hop Count = 0; //for RNs beginning of round.
9.   Identify source RN as a start node for current round.
10.   List all neighbor RN indices from source RN
11.   **If** source RN has one-hop, send directly to GCN.
12.   **Else If** there is least one RN connected with this RN
13.   **For** each source RN index 'j' do
14. $f(n_j)$ = Distance (Neighbor RN to GCN)
15. $g(n_j)$ = Energy (Neighbor RN's energy)
16. $h(n_j) = min\ (f(n_j))$
17. $i(n_j)$ = InitialEnergy (RN's initial Energy)
18.   **End**
19.   **If** $g(h(n_j)) > i\ (h(n_j)) * 50\%$
20.   Chosen RN Index = $g\ (h(n_j))$
21.   **End**
22.   **End**
23.   **End**
24.   **Else**
25.   There is no source RN
26.   **End**

27. **If** RN's energy when connected with GCN < 0,
28. **Then** disconnect from path
28. **End**
29. Update neighbor energy information of source RN
30. **Termination Criteria**
31. $P(l(z, X) \le r) = 0$ in Equation 6.5 is equal to 0.
32. **End**
33. **Return** $g(h(nj))$

In the above algorithm, elements of cognition in the utilized *cognition process* form the two main constituents of our proposed approach. The elements that help in implementing cognition in the cognitive nodes are: *reasoning* and *learning* elements.

6.4.1 Learning

Learning is used in our CEEA approach in order to determine the most appropriate paths toward the GCN that satisfy the IoT network requirements. This cognition element uses a direction-based heuristic to determine the data delivery path through RNs that lie in the direction of the GCN. Hence, each time the *cognition process* has to choose the next hop, the direction-based heuristic eliminates RNs that increase the distance between the current RN and GCN. Knowledge of the positions of the CRN and its one-hop RNs is used by the heuristic to determine the set of such RNs, which we call forward-hop-RNs. Thus the forward-hop-RNs of a CRN identified by the direction-heuristic is constituted by those RNs that reduce the distance between the CRN and the GCN. This information is stored in the CRN for use in the next transmission rounds. Thus the direction-based heuristic, along with feedback from the network about the chosen paths helps the *cognition process* to learn data delivery paths to the sink, as the network topology changes.

Example 6.1: Assume S_1 and S_2 have data to be sent to destination nodes D_1 and D_2. R_n are all the available relays toward the destination. Out of these relays, it is determined that R_5 as shown in Figure 6.1 has the lowest link outage probability to D_1 and D_2. Therefore, S_1 initiates routing data to R_5. Meanwhile, S_2 also forward a high traffic

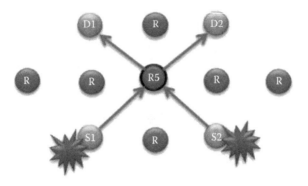

Figure 6.1 Classical routing in a sensor network.

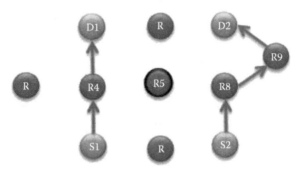

Figure 6.2 Cognitive routing in IoT-networks.

of data to R_5 (depicted by solid paths in Figure 6.1). When multiple source nodes start routing their data to R_5 as well, the route to R_5 may get congested. A cognitive network with *learning* capabilities will be able to identify the congestion at R_5 (by observing the decrease in throughput). Sharing this observation with neighboring nodes, the cognitive IoT network would be able to respond to the congestion proactively, by routing the data through a different path involving nodes R_4, R_8, and R_9, as shown in Figure 6.2.

6.4.2 Reasoning

In the CEEA approach, we assume a modified version of the analytic hierarchy process (AHP) [18] for implementing the reasoning element of cognition in the IoT. AHP supports multiple-criteria decision making while choosing the data path. For example, if we have a delay-sensitive data, the node which provides the lowest delay will be chosen even though it might degrade other metrics such as the network energy or throughput. If two next hops guarantee the same delay then the next attribute to compare will be energy, and then, throughput, assuming that energy is the next desired attribute in the IoT-network. AHP provides a method for pairwise comparison of each of the attributes and helps to choose the node that can provide the best network performance on the long run. The following subsequent example has more details on the utilized AHP. While AHP calculations help in deciding the next hop, it also helps in planning for future actions. The *cognition process* enables the CEEA approach to maintain the calculated values of the IoT-network attributes, which can be used in future transmission rounds. Hence, these values are not necessarily calculated at every transmission round.

Example 6.2: Assume a three-level hierarchy in the AHP: *Goal, Criteria,* and *Alternatives* as shown in Figure 6.3. A fundamental scale for pairwise comparisons is then used to set priorities for the IoT-network attributes/criteria at the CNs. Given the very limited energy constraint in IoT, we would assign the highest priority to energy, followed by *reliability* and then *throughput*. We tabulate the relative priorities of these attributes using pairwise comparison in [36] and generate Table 6.1. From Table 6.1 we generate Table 6.2. Then, we apply the following steps:

1. Represent the values of Table 6.2 in the matrix form

$$A = \begin{bmatrix} 1 & 4 & 6 \\ 1/4 & 1 & 3 \\ 1/6 & 1/3 & 1 \end{bmatrix}.$$

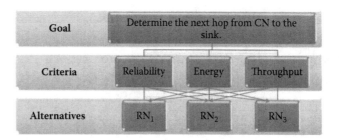

Figure 6.3 The AHP hierarchy.

Table 6.1   Pairwise Comparison of the IoT-Network Attributes

Energy (Kj)	4	Reliability	1
Energy (Kj)	6	Throughput (Mbps)	1
Reliability	3	Throughput (Mbps)	1

Table 6.2   AHP for QoI Attributes v/s Goal

GOAL – BEST ATTRIBUTE	ENERGY (Kj)	RELIABILITY	THROUGHPUT (Mbps)	RELATIVE PRIORITIES OF THE ATTRIBUTES
Energy (Kj)	1	4	6	0.691
Reliability	1/4	1	3	0.2176
Throughput (Mbps)	1/6	1/3	1	0.0914

Table 6.3   AHP Evaluating the Overall Priorities for All Possible RNs

BEST CANDIDATE FOR NEXT HOP RN_X	PRIORITY WITH RESPECT TO			
	ENERGY (Kj)	RELIABILITY	THROUGHPUT (Mbps)	GOAL
RN_1	0.252	0.015	0.101	0.375
RN_2	0.2	0.018	0.11	0.329
RN_3	0.164	0.019	0.116	0.296

2. Compute the Eigen vector of the matrix A.
3. Isolate the absolute, real values of the Eigen vector.
4. Compute the relative priority values.

Note that our goal is to find the best next hop, which provides the highest value for a specific attribute, as shown in Table 6.3.

Algorithm 6.1b AHP Analysis for Path Selection in *Cognition Process*

1. Function AHP (priorities of the attributes P)
2. Input
3. P: End user defined priorities on the attributes for requested data
4. Output
5. RN_x: Forward-hop $RN_x \in \{RN_1 \ldots RN_n\}$ with best P
6. Begin
7. Initialize: priority matrix for traffic type; Success=0;

8. While $P(l(z, X) \leq r) > 0$ in Equation 6.5

9.   *AHP_analysis*(Next hop RNs v/s attributes)

10.   Next hop $RN = RN_x$

11.   Transmit data to next hop RN

12.   **If** (next hop = GCN)

13.   Success=1;

14.   **Else**

15.   *Choose next hop RN*

16.   goto step 8

17.   **End**

18.   **If** *(Success==0)*

19.   GCN Retransmits request

20.   **End**

21. End

If energy consumption is measured as a function of the number of events taking place before the data packet arrives to the sink, the hop count can be used to approximate the energy cost. Accordingly, the modified AHP steps in prioritizing the IoT network attributes and identifying the best next hop are described in Algorithm 6.1b.

6.5 Performance Evaluation

In this section, we evaluate the performance of the proposed CEEA. We use MAEB, GEAR, ReInForM, and LinGo algorithms as baseline evaluation algorithms. Based on the aforementioned system models, we summarize these four baselines' categories as follows.

Geographic and Energy-Aware Routing (GEAR)

     In this approach, sensor nodes must have a hardware component for positioning such as a GPS unit or a localization system. GEAR routing protocol is used to improve the efficiency in terms of energy consumption via forwarding queries to targeted regions. The forwarding scheme operates at two phases: *Setup* phase and *operation* phase. Setup phase is designed to assist sensor nodes in measuring their distances from the anchors. In the operation phase, a source sensor node incorporates its location information in a packet header. A receiving node checks its location, the destination location and source location for either forwarding or dropping the received packet.

Reliable Information Forwarding Using Multiple Paths (ReInForM)

ReInForM employs a probabilistic flooding procedure to deliver information-aware packets at a predetermined priority level. This leads to more reliable routing protocol at a proportional data delivery cost. The routing mechanism is based on local knowledge of network conditions, such as channel error, and hop-count to sink.

Movement-Aided Energy-Balance (MAEB)

MEAB has been chosen as a baseline due to movement and energy consideration. It has a neighbor discovery procedure which is conducted by the network cluster heads. They send their data packet to the Gateway following a forwarding rule, in which the distance and velocity to the Gateway and the remaining energy is recorded to select the neighbor cluster heads on the route towards the Gateway.

Link Quality and Geographical Beaconless OR Protocol (LinGo)

LinGo introduces a different progress calculation approach compared to the aforementioned ones [37]. It takes into account both the progress of a given forwarding node towards the destination with respect to the last-hop, as well as the radio range. In this way, LinGO reduces the number of required hops on a data towards the destination node.

Cognitive Networking with Opportunistic Routing (CNOR)

CNOR protocol [38] is designed mainly for scalable WSNs, and tries to combine the advantages of opportunistic routing and opportunistic spectrum access. It is a reactive routing protocol since it discovers routes only when desired. An explicit route discovery process takes place only when it is needed. The destination node of the network begins the route discovery process and this process ends when a routing path has been established while a maintenance procedure preserves it until the path is no longer available or desired.

As the network scalability is increased, CNOR tends to discover more paths leading to the increase of the network performance.

Energy-Aware Routing for Cognitive Networks (EARCN)

EARCN scheme [39] associates the backward difference traffic moments with the sleep-time duration to tune the activity durations of a node for achieving optimal energy conservation and alleviating the uncontrolled energy consumption in wireless devices. It provides efficient cognitive routing protocol in terms of maximum energy conservation, maximum-possible routing paths establishments, and minimum delays, while utilizing secondary communication nodes (e.g., CRNs).

Resilient IoT for Dynamic Sensor Networks (RIDSN)

RIDSN extends AODV protocol to match needs of cognitive ad hoc networks in the IoT paradigm. It dissects the study of any IoT nodal capacity to its "connected" components, and empowers dynamic associativity between things to serve varying functional requirements and levels [40]. More importantly, critical resources in the network will be shared within their neighborhoods. Thus network lifetime will relate to functional cliques of dynamic IoT nodes, rather than individual networks.

6.5.1 Performance Metrics and Parameters

To compare the performance of these five schemes, the following four performance metrics are used.

- Average Delay: Is measured in msec and is defined as the average amount of time required to deliver a data unit to the destination.
- Idle time: This metric reflects the ratio of idle time every node spends while just waiting to forward a message. It is measured in μsec.
- Throughput: Is set here as a quality measure. It is the average percentage of transmitted data packets that succeed in reaching the destination reflecting the effect of node heterogeneity and delay in IoT setups over the utilized data delivery approach.

- Average Price: This metric is used to observe the influence of the utilized data delivery approach on the overall price to deliver a data unit from source to destination on average. The price charged by each node n_i as p_i.

$$p_i = \gamma_i * \left[\frac{E_{Tx}(D_k, n_j) + E_{Rx}(D_k)}{e_i} + \acute{\pi}_\iota + \acute{u}_\iota \right] \qquad (6.6)$$

where $\acute{u}_\iota$ is the available buffer space at node i, and $\acute{\pi}_\iota$ is the power amount to be consumed per packet processing at node i. $E_{Tx}(D_k, n_j)$ and $E_{Rx}(D_k)$ are the mounts of energy used to transmit a data packet D_k from node i to j and receive a data packet D_k at node i, respectively. And e_i is the instantaneous available energy per node i.

Meanwhile, the three data delivery performance is assessed using the following three parameters:

- The size of the network in terms of total node count. This reflects the application's complexity and the scalability of the exploited routing scheme. Knowing that larger node count in a data path raises the risk of node failure and, hence, dropped packets.
- Energy: Average energy consumption rate per data unit ($\acute{\pi}_\iota$) as an indicator of the network power saving. This metric is measured in kilojoule. This parameter is thus assumed to be always greater than zero in the following simulation results.
- PNF (%): It is the probability of a physical damage or a battery depletion for the deployed sensor node due to a disaster harsh-operational conditions. This parameter is chosen to reflect the impact in case of disaster scenarios or fragmented networks in IoT.
- Cost (γ_i) to observe the influence of the charged price rate over the utilized data delivery approach. It is a pricing factor for each node in the IoT measured in dollar/byte. This is can be set as a flat rate per number of bytes transmitted, where setting it to a relatively high value would diminish the chances of n_i to be selected for relaying the data packet D_k.

6.5.2 Experimental Setup

In order to limit our search space, we assume a virtual grid, where SNs are placed on the grid vertices. We assume up to 1500 total SNs communicate with one GCN via 36 RNs. We used NS3 as simulation tool for this purpose. The simulation is processed in three platforms, which are Windows, Linux, and OSX, for validation purposes. We executed our simulation 100 times for each experiment and plotted the average results. More details about our simulation are summarized in Table 6.4.

6.5.3 Simulation Results

In Figure 6.4, the experienced delay in delivering data packet is plotted against the size of the network for the different simulated algorithms.

Table 6.4 Simulation Parameters and Values

PARAMETER	VALUE
Targeted area	1000 m × 1000 m
Number of nodes	SNs: 350, RNs: 36, GCN: 1
Communication Range	SN: *142 m*, RN: *300 m*, GCN: *500 m*
Initial Energy	SN: *31104J*, RN: *110160J*, GCN: *Unlimited*
Energy Consumption	SN and RN (Receiving): *31.2 uJ/bit* SN and RN (Transmitting): *53.8 uJ/bit*

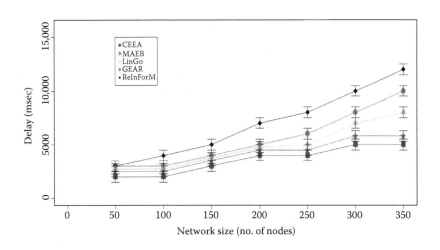

Figure 6.4 Delay vs. the number of nodes in an IoT network.

We observe that ReInForM has the highest delay, while CEEA has the lowest delay as the number of nodes increases. Therefore, we can say that CEEA is more delay-tolerant in comparison to all of the sampled algorithms. We also observe that there is a monotonic increase in delay for ReInForM algorithm, while MAEB has a slightly higher delay than CEEA with a constant difference at every node. For CEEA and MAEB, we observe a steep increase between 100 and 200 nodes while RelnForM has its steepest slop between 150 and 200. For LinGo and GEAR we observe a fairly continues increase in delay as the number of nodes increase since they are more dependent on the network nodes' geolocations.

Figure 6.5 shows the experienced network throughput versus the number of nodes for the sampled algorithms. We can observe that there is a general increase in throughput of the sampled algorithms as the size of the network increases. MAEB, LinGo, and GEAR have the same throughput until the size of the network is about 150 nodes, after which MAEB gives a higher throughput. Also, it's worth remarking here that LinGO adds redundant packets in order to increase the packet delivery probability while experiencing link error periods. This leads to significant increment in the overall throughput in comparison to GEAR and ReInForM methods. From the graph, we can also observe that in all instances, CEEA has a higher throughput than the others do, and ReInForM has the lowest throughput. And hence, we can conclude that CEEA has a better throughput as

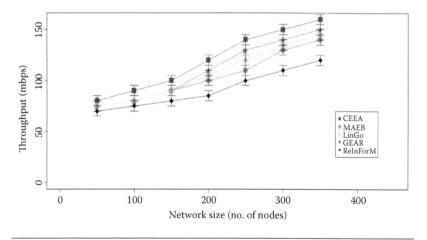

Figure 6.5   Throughput vs. the network size.

the network size increases compared to the sampled algorithms. This can be returned to the efficient retransmission approach in CEEA algorithm in comparison to other approaches in the literature. This makes it also the most scalable approach for the next generation IoT networks where the connected network nodes are dramatically increasing a day after a day.

Plotted curves in Figure 6.6 show the average consumed energy against throughput for the different examined algorithms. We notice that there is almost a liner increase in energy consumption while applying the RelnForM approach as the network throughput increases, while CEEA, GEAR, LinGo, and MAEB forms a concave-like curves. We also observe that for every amount of energy consumed, RelnForM has the lowest throughput. On the other hand, CEEA has the highest throughput for the same amount of energy. For this reason, we can conclude that CEEA is the most efficient algorithm in terms of energy consumption compared to the sampled ones. Moreover, we notice that when the energy budget is greater than or equal to 60 kilojoules, the network throughput is saturated due to other design factors such as the network size and cost factor (γ_i).

Figure 6.7 shows the average charged cost (γ_i) per network node i against the experienced data delivery delay for all the simulated approaches. From this figure we can observe that with the increase in gamma, there is a general decrease in delay time for all the sampled schemes. Which is an expected network behavior as the flat rate

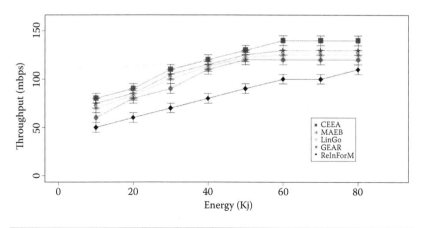

Figure 6.6 Throughput vs. the energy consumption.

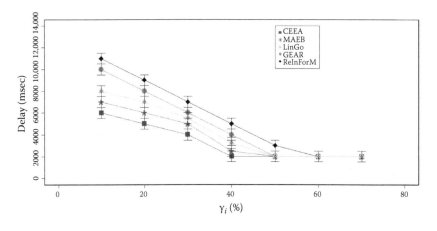

Figure 6.7   Delay vs. the average γ_i rate percentage.

charge increases per node. We also notice that all the schemes reach a certain threshold where the delay becomes constant at 2000 milliseconds. CEEA is the first to get to the threshold when γ_i equals to 40, while ReInForM is the last when γ_i equals to 60. GEAR and MAEB reaches the threshold at γ_i equals to 50. Consequently, we can say that CEEA is the most cost-effective scheme since it has the lowest delay time with the lowest γ_i.

In Figure 6.8, average idle time is compared under varying total count of network nodes. As the network size, or the number of SNs increases, there is a general increase in the average idle time. However, we observe that CEEA has the lowest idle time compared to other baselines. From Figure 6.8 we can also deduce that after a

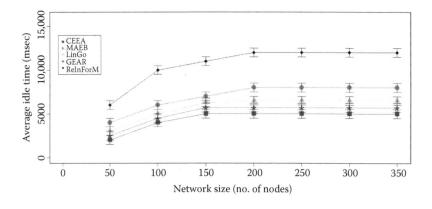

Figure 6.8   Average idle time vs. the network size.

network size of 150 nodes, the average idle time of CEEA remains constant, which means it is not affected by the number of nodes. MAEB has a slightly higher idle time than CEEA, whose difference to CEEA remains constant as the network increases in size. GEAR and ReInForM have an increasing idle time until 200 nodes, and then stay in a steady state. Therefore, we can conclude that CEEA is most efficient compared to the sampled baseline algorithms.

Figure 6.9 depicts the network size against the average price of all the schemes. From the figure, we can observe that ReInForM has the highest average price. On the other hand, the CEEA approach has the lowest average price under all varying node counts. When the number of nodes reaches 250, the ReInForM approach has a constant and fixed average price. On the contrary, after a network size of 250 nodes, we observe a sharp and linear increase on GEAR and LinGo. Meanwhile, MAEB is the second-most scheme that has the lowest average price after the CEEA approach. The achieved price curve of MAEB closely follows that of the CEEA. However, it is still worse than the CEEA. Therefore, CEEA has the best performance in terms of average price as well under all experimented network sizes. The reason is that GEAR, LinGO, and MAEB approaches add redundant packets in order to increase the packet delivery probability while experiencing link error periods. This leads to significant increment in the overall price.

In Figure 6.10, the y-axis represents energy levels of specified RNs and the x-axis represents the specified RNs and the algorithm types.

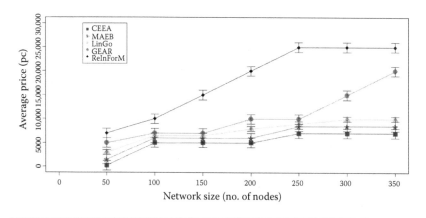

Figure 6.9 Average price vs. the network size.

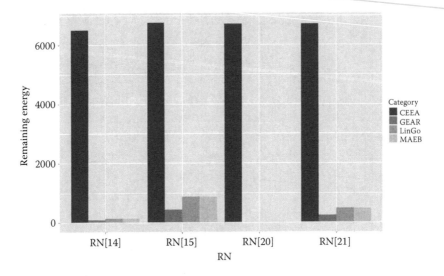

Figure 6.10   Comparison of One-hop RNs' Energy Level.

The reason why RN_{14}, RN_{15}, RN_{20}, and RN_{21} are selected is because these RNs have bidirectional connections with the GCN. To transmit a packet to the GCN, one of these RNs must be used. We compare these RNs energy levels against the ReInForM, GEAR, LinGo, and MAEB algorithms, since they are the most energy-efficient ones. Obviously, GEAR has the worst performance in this figure due to a fairness problem in this algorithm while relaying towards the sink node. Although the energy level of RN_{14}, RN_{15}, RN_{20}, and RN_{21} are better for LinGo and MAEB, these RNs' energy levels are significantly outperformed by the CEEA approach. Thus, CEEA increases the network lifetime and it is better for energy-saving. Furthermore, when we compare these algorithms in terms of the number of transmission rounds, it can be clearly observed from the simulation results in Figure 6.11 that CEEA outperforms GEAR, LinGo, and MAEB. Notably, the more savings in terms of remaining energy shown in Figure 6.10 by applying the CEEA approach have led to prolonged network lifetime in Figure 6.11.

Moreover, we examined the four routing approaches; CEEA, LinGo, GEAR, and MAEB in terms of the average delay impacts (Figure 6.12) while considering disaster scenarios or fragmented network, where failure of a critical node partitions the network into disjoint segments. Based on Figure 6.12, we notice a severe effect on the

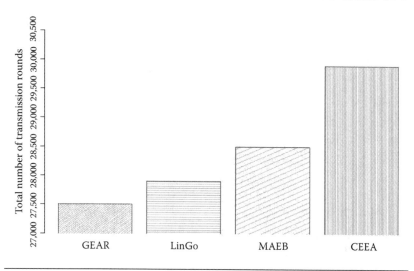

Figure 6.11 Comparison of the four data delivery techniques based on total number of transmissions.

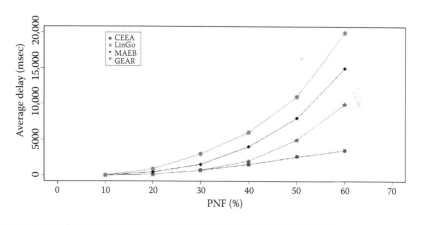

Figure 6.12 Average delay versus the probability of node failure in the network.

average delay while the probability of node failure (PNF) is increasing. We notice that all approaches are experiencing an exponential increase in the experienced delay as the network becomes disconnected. However, using the proposed CEEA approach, the increment is going linear, which can be a very desirable feature in IoT while experiencing harsh operational conditions and severe mobility effects.

The good performance achieved by the CEEA approach in this chapter can be returned mainly to the utilized cognitive elements that help a lot in disaster scenarios. In fact, the proposed CRNs make use of the received feedback about the utilized channel condition and modulation rate to determine the sleep time of each node. This concept has been emphasized more in Figure 6.7 while assuming realistic channel conditions as summarized in Table 6.5 for energy consumption of the relay node in four modes: Sleep mode, receive mode, active mode (ready to transmit but not transmitting), and transmission mode. Figure 6.13 displays the mean node lifetime in the network using adaptive modulation and adaptive sleep (AMS) versus adaptive modulation (AM) only. In case of AM, the modulation level (parameter M in M-QAM modulation) is chosen for each packet according to channel condition (i.e., SNR). This case assumes no cognition and

Table 6.5   Node Parameters Used in Simulation Based on [29]

PARAMETER	VALUE
Current consumption in sleep mode: I_{sleep}	1 μA
Current consumption in receive mode: I_{rx}	20 mA
Current consumption in active mode: I_{ac}	100 mA
Current consumption while transmitting	120 mA
Traffic intensity	90 percent
Log–Normal Shadowing variance (σ)	0, 2 dB, 4 dB, or 6 dB
BER required (QoS)	10^{-4}
RF Bandwidth used	200 kHz

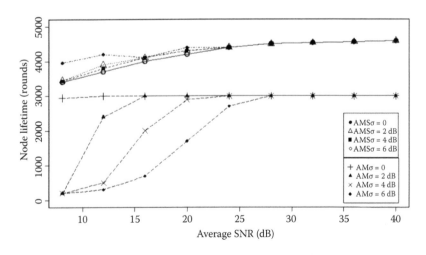

Figure 6.13   Node lifetime using adaptive modulation (AM) vs. adaptive modulation and adaptive sleep (AMS).

the sleep time is predetermined independently from the user requests or any changes to application requirements. Meanwhile, the same figure shows the average node life time using an adaptive modulation scheme combined with a scheduled sleep via the AMS mechanism that adapt based on the CRNs feedback. In other words, cognition here is employed at the MAC and PHY layers and sleep times are scheduled according to channel conditions and bit rates. In both cases high traffic patterns (mean packet arrival rate 90 percent) are assumed and simulated using Poisson distributions and log–normal shadowing where shadowing variance takes values from 0 (no shadowing) to 6 dB. The figure shows that the cognitive approach significantly outperforms the noncognitive one in terms of the average node lifetime. This is because M-QAM modulation, when carefully chosen, will require less transmission time, and thus the cognitive system exploits this information to modify the sleep time accordingly. The improvement made by the cognitive approach is higher for both high traffic intensity and severe channel conditions (i.e., low SNR and high shadowing variance).

Considering the fact that the proposed CEEA technique is being developed for the use in a dynamic IoT environment, it is desirable to continuously improve the success rate of the data delivery in order to improve the in-network user experience. Being surer of getting a response back from the network for each of the queries sent out can significantly improve the levels of the user's satisfaction with the network. However, the aforementioned benchmarks were not able to efficiently learn and adapt to dynamic network changes such as node failures and remaining energy in the IoT paradigm when we compared to the proposed CEEA data delivery approach. Thus, more cognitive benchmarks, such as CNOR, RIDSN, and EARCN, have been further investigated in this study against the proposed CEEA approach.

Figures 6.14 and 6.15 compare the average, minimum, and maximum values of success rates and failure rates, respectively, for all of the four approaches. We can see that the CEEA approach has the highest average success rate of 88 percent, and its worst-case failure rate is only 3 percent more than CNOR. Since it is more desirable to have a higher success rate in smart IoT applications, we further compare the performance of CEEA and CNOR techniques in terms of

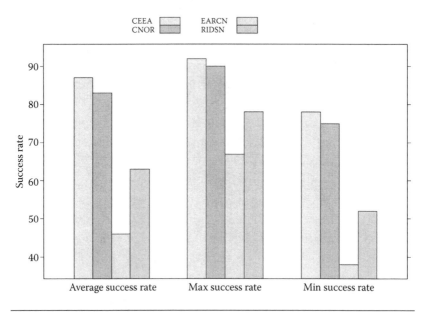

Figure 6.14 Comparison of success rates.

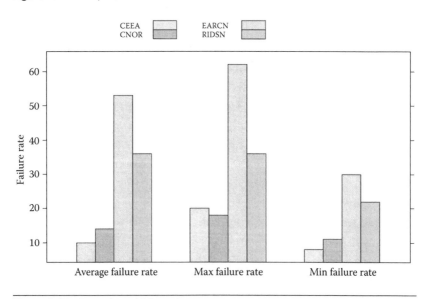

Figure 6.15 Comparison of failure rates.

their effective-QoI (eQoI) as observed at the sink to identify the best approach of the two, where the eQoI is the heuristics estimate of the QoI associated with data delivered to the sink at the end of a successful transmission round. In other words, it is an estimate for the value of QoI at the last hop that delivers directly to the sink.

Figure 6.16 shows the result of the comparison of the eQoI values for CEEA and CNOR, with RIDSN, which doesn't use any kind of learning at the CRNs. In general, we observe that using some form of learning at the CRNs improves the eQoI of the data delivered to the sink. Among the learning techniques, we observe that CEEA performs the best in terms of consistently delivering data with higher eQoI at the sink, even toward the end of the network's lifetime. Now, this eQoI is the hop-by-hop value of QoI associated with the data delivered to the sink with respect to energy, reliability, and throughput. In addition, the cumulative delay in receiving a response from the network for an initiated request by the sink node is reflected by the number of hops taken along the path from the source to sink.

Consequently, we compare the hop count against the network life-time for CEEA, CNOR, and RIDSN, and the results are as shown in Figure 6.17. The observation made from the compared techniques is the spike in hop count seen towards the end of the network's lifetime. This is because more number of nodes are lost due to node deaths as the simulations progress, making it increasingly difficult for the remaining alive nodes to find paths to deliver data to the sink. This search for alternate paths leads to an increase in the hop count. However, one of the marked differences between CEEA and CNOR is that CNOR has a constant hop-count of two till about 300 trans-mission rounds. This is because it starts out with exploiting its knowl-edge of paths through RNs that are one-hop away from the sink. But CEEA starts with exploring paths and eventually learns the

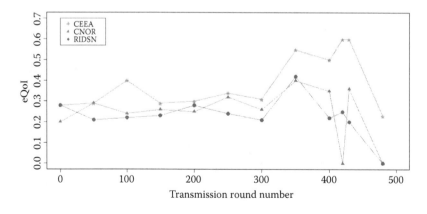

Figure 6.16 Comparison of eQoI as observed at the sink over the network lifetime.

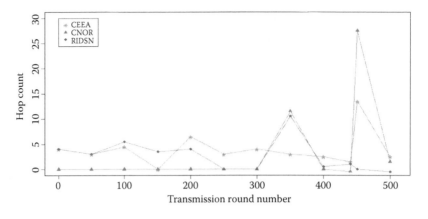

Figure 6.17   Comparison of the average hop counts in delivering data from a source to the sink.

network connections that helps to reduce the worst case hop-count towards the end of the network's lifetime. This difference in strategies accounts for the slightly lower network lifetime of CEEA when compared with CNOR, due to higher energy consumed in exploring the paths. However, when averaged over the entire network lifetime, CEEA is on average, two hops more expensive than CNOR, which rewards its higher average rate of successful data delivery and eQoI at the sink.

6.6 Conclusion

In this chapter, we investigated routing techniques for the IoT paradigm in terms of energy consumption, cost, and delay, while experiencing harsh operational conditions and sever energy limitations. We proposed a novel approach for sensor networks in IoT, called CEEA. We found that CEEA can save a considerable amount of energy. Moreover, we showed how the data delivery price can be affected by the network size for varying energy-based routing approaches. Furthermore, the CEEA approach was compared to other cognitive routing approaches in ad hoc networks. It was able to provide a 40-percent improvement in average data delivery success rate when compared with the RIDSN. CNOR approach performed equally well in terms of the data delivery success rate, but performed slightly better than CEEA in terms of the energy consumed. On the other hand, the CEEA approach is a better choice when the application requires a higher eQoI at the sink, and higher best-case success rate. Consequently, CEEA approach is recommended for disaster-inspired applications which could provide data for other approaches to identify disasters in real-time [41–44] and it outperforms key other baseline approaches in terms of transmission and energy consumption.

References

1. F. Al-Turjman, H. Hassanein, and M. Ibnkahla, "Efficient Deployment of Wireless Sensor Networks Targeting Environment Monitoring Applications," *Elsevier Computer Communications*, vol. 36, no. 2, pp. 135–148, 2013.

2. M. Hijji, S. Amin, R. Iqbal, and W. Harrop, "The Significance of Using Expert System to Assess the Preparedness of Training Capabilities against Different Flash Flood Scenarios," *Lec Notes on Software Engineering*, vol. 3, no. 3, pp. 214–219, 2015.

3. F. Al-Turjman, H. Hassanein, W. Alsalih, and M. Ibnkahla, "Optimized Relay Placement for Wireless Sensor Networks Federation in Environmental Applications," *Wiley Wireless Communication & Mobile Computing*, vol. 11, no. 12, pp. 1677–1688, 2011.

4. G. Tolle, J. Polastre, R. Szewczyk, and D. Culler, "A Macroscope in the Redwoods," in *Proc. of the ACM Conf. on Embedded Networked Sensor Systems*, San Diego, pp. 51–63, 2005.

5. M. Z. Hasan, F. Al-Turjman, and H. Al-Rizzo, "Optimized Multi-Constrained Quality-of-Service Multipath Routing Approach for Multimedia Sensor Networks," *IEEE Sensors*, 2017. DOI: 10.1109/JSEN.2017.2665499.

6. C. Dorazio, K.-K. R. Choo, and L. T. Yang, "Data Exfiltration from Internet of Things Devices: iOS Devices as Case Studies," *IEEE Internet of Things*, vol. pp, no. 99, 2016.

7. Y. Yang, H. Cai, Z. Wei, H. Lu, K. Kwang, and K.-K R. Choo, "Towards Lightweight Anonymous Entity Authentication for IoT Applications," in *Proc. of the Australasian Conference on Information Security and Privacy*, Melbourne, Australia, pp. 265–280, 2016.

8. N. Cahyani, B. Martini, K.-K. R. Choo, and M. Al-Azhar, "Forensic Data Acquisition from Cloud-of-Things Devices: Windows Smartphones as a Case Study," *Wiley Concurrency and Computation: Practice and Experience*, 2016. DOI: 10.1002/cpe.3855.

9. L. A. Al-Fuqaha, M. Guizani, M. Mohammadi, M. Aledhari, M. Ayyash, "Internet of Things: A Survey on Enabling Technologies, Protocols, and Applications," *IEEE Communications Surveys & Tutorials*, vol. 17, no. 4, pp. 2347–2376, 2015.

10. S. Ali and S. Madani, "Distributed Efficient Multi Hop Clustering Protocol for Mobile Sensor Networks," *International Arab Journal of Information Technology*, vol. 8, no. 3, pp. 302–309, July 2011.

11. T. M. Cover and J. A. Thomas, *Elements of Information Theory*, Wiley Press, Malden, MA, 1991.

12. S. A. Nikolidakis, D. Kandris, D. D. Vergados, and C. Douligeris, "Energy Efficient Routing in Wireless Sensor Networks Through Balanced Clustering," *MDPI Algorithms*, vol. 6, no. 1, pp. 6, 29–42, 2013.

13. A. Aburumman, and K.-K. R. Choo, "A Domain-Based Multi-cluster SIP Solution for Mobile Ad Hoc Network," in *Proc. of the Int. Conf. on Security and Privacy in Communication Systems*, Beijing, China, pp. 267–281, 2014.

14. T. Maniak, C. Jayne, R. Iqbal, and F. Doctor, "Automated Sound Signalling Device Quality Assurance Tool for Embedded Industrial Control Applications," in *Proc. of the IEEE Int. Conf. on Systems, Man, and Cybernetics*, Manchester, UK, pp. 600–611, 2013.

15. F. Doctor, C.-H. Syue, J.-H. Shieh, and R. Iqbal, "Type-2 Fuzzy Sets Applied to Multivariable Self-Organizing Fuzzy Logic Controllers for Regulating Anesthesia," *Journal of Applied Soft Computing*, vol. 38, no. 1, pp. 872–889, 2016.

16. F. Al-Turjman, "Cognition in Information-Centric Sensor Networks for IoT Applications: An Overview," *Annals of Telecommunications*, vol. 72, no. 1, pp. 1–16, 2016.

17. N. Kumar, R. Iqbal, S. Mistra, and J. Rodrigues, "Bayesian Coalition Game for Contention Aware Reliable Data Forwarding in Vehicular Mobile Cloud," *Elsevier Future Generation Computer Systems*, vol. 48, no. 1, pp. 60–72, 2014.

18. H. Fang, L. Xu, and K.-K. R. Choo, "Stackelberg Game-Based Relay Selection for Physical Layer Security and Energy Efficiency Enhancement in Cognitive Radio Networks," *Elsevier Applied Mathematics and Computation*, vol. 296, no. 1, pp. 153–167, 2017.

19. H. Fang, L. Xu, J. Li, and K.-K. R. Choo, "An Adaptive Trust-Stackelberg Game Model for Security and Energy Efficiency in Dynamic Cognitive Radio Networks," *Elsevier Computer Communications*, 2016. DOI: 10.1016/j.comcom.2016.11.012.

20. A. Aburumman, W. Seo, C. Esposito, A. Castiglione, R. Islam, and K.-K. R. Choo, "A Secure and Resilient Cross-Domain SIP Solution for MANETs using Dynamic Clustering and Joint Spatial and Temporal Redundancy," *Wiley Concurrency and Computation: Practice and Experience*, 2016. DOI: 10.1002/cpe.3978.

21. B. Deb, S. Bhatnagar, and B. Nath, "ReInForM: Reliable Information Forwarding Using Multiple Paths in Sensor Networks," in *Proc. of the IEEE Conf. on Local Computer Networks*, Bonn, Germany, pp. 406–415, 2003.

22. T. El Salti, D. Stacy, N. Nasir, and F. Al-Turjman, "Packet Delivery Significance and Metrics Improvements in Protocols for 3-D Routing in Wireless Sensor Networks," in *Proc. of the Int. Wireless Communications and Mobile Computing Conf.*, Nicosia, Cyprus, pp. 1130–1135, 2014.

23. S. Mahmud, R. Iqbal, and F. Doctor, "Cloud Enabled Data Analytics and Visualization Framework for Health-Shocks Prediction," *Elsevier Future Generation Computer Systems*, vol. 65, no. 1, pp. 169–181, 2016.
24. G. Singh and F. Al-Turjman, "A Data Delivery Framework for Cognitive Information-Centric Sensor Networks in Smart Outdoor Monitoring," *Elsevier Computer Communications*, vol. 74, no. 1, pp. 38–51, 2016.
25. L. Villalba, A. Orozco, A. Cabrera, and C. Abbas, "Routing Protocols in Wireless Sensor Networks," *MDPI Sensors*, vol. 9, no. 1, pp. 8399–8421, 2009.
26. M. Z. Hasan, H. Al-Rizzo, and F. Al-Turjman, "A Survey on Multipath Routing Protocols for QoS Assurances in Real-Time Multimedia Wireless Sensor Networks," *IEEE Communications Surveys and Tutorials*, 2017. DOI: 10.1109/COMST.2017.2661201.
27. A. Tsioliaridou, C. Liaskos, S. Ioannidis, and A. Pitsillides, "CORONA: A Coordinate and Routing system for Nanonetworks," in *Proc. of the ACM Int. Conf. on Nanoscale Computing and Communication*, Boston, MA, pp. 1–6, 2015.
28. S. Oteafy, "A Framework for Heterogeneous Sensing in Big Sensed Data," in *Proc. of the IEEE Global Communications Conf.*, Washington, DC, pp. 1–6, 2016.
29. F. Saghezchi, A. Radwan, and J. Rodriguez, "Energy-Aware Relay Selection in Cooperative Wireless Networks: An Assignment Game Approach," *Elsevier Ad Hoc Networks*, vol. 56, no. 1, pp. 96–108, 2017.
30. B. Baranidharan and B. Santhi, "An Evolutionary Approach to Improve the Life Time of the Wireless Sensor Networks," *Theoretical and Applied Information Technology*, vol. 33, no. 2, pp. 177–183, 2011.
31. F. Al-Turjman, "Hybrid Approach for Mobile Couriers Election in Smart-cities," in *Proc. of the IEEE Local Computer Networks*, Dubai, UAE, pp. 507–510, 2016.
32. A. Al-Hourani and S. Kandeepan, "Cognitive Relay Nodes for Airborne LTE Emergency Networks," in *Proc. of the IEEE Int. Conf. on Signal Processing and Communication Systems*, Gold Coast, Australia, pp. 1–9, 2013.
33. J. M. Jornet and I. F. Akyildiz, "Channel Modeling and Capacity Analysis for Electromagnetic Wireless Nanonetworks in the Terahertz Band," *IEEE Trans. Wireless Commun.*, vol. 10, no. 10, pp. 3211–3221, 2011.
34. M. Z. Hasan, F. Al-Turjman, and H. Al-Rizzo, "Evaluation of a Duty-Cycled Protocol for TDMA-Based Wireless Sensor Networks," in *Proc. of the IEEE Int. Wireless Communications and Mobile Computing Conf.*, Paphos, Cyprus, pp. 964–969, 2016.
35. H. Yu, B. Ng, and W. K. G. Seah, "Forwarding Schemes for EM-Based Wireless Nanosensor Networks in the Terahertz Band," in *Proc. of the 2nd Annual ACM International Conference on Nanoscale Computing and Communication*, Boston, MA, USA, pp. 1–6, 2015.
36. G. Singh and F. Al-Turjman, "Learning Data Delivery Paths in QoI-Aware Information-Centric Sensor Networks," *IEEE Internet of Things*, vol. 3, no. 4, pp. 572–580, 2016.

37. D. Rosário, Z. Zhao, A. Santos, T. Braun, and E. Cerqueira, "A Beaconless Opportunistic Routing Based on a Cross-Layer Approach for Efficient Video Dissemination in Mobile Multimedia IoT Applications," *Elsevier Computer Communications,* vol. 45, no. 1, pp. 21–31, 2014.

38. P. Spachos and D. Hantzinakos, "Scalable Dynamic Routing Protocol for Cognitive Radio Sensor Networks," *IEEE Sensors,* vol. 28, no. 15, pp. 4038–4052, 2016.

39. A. Bourdena, C. Mavromoustakis, G. Kormentzas, E. Pallis, G. Mastorakis, and M. Yassein, "A Resource Intensive Traffic-Aware Scheme Using Energy-Aware Routing in Cognitive Radio Networks," *Elsevier Future Generation Computer Systems,* vol. 39, no. 1, pp. 16–28, 2014.

40. S. Oteafy and H. Hassanein, "Resilient IoT Architectures over Dynamic Sensor Networks with Adaptive Components," *IEEE Internet of Things,* vol. 4, no. 2, pp. 474–483, 2016.

41. Z. Xu, H. Zhang, C. Hu, L. Mei, J. Xuan, K.-K. R. Choo, V. Sugumaran, and Y. Zhu, "Building Knowledge Base of Urban Emergency Events Based on Crowdsourcing of Social Media," *Wiley Concurrency and Computation: Practice and Experience,* vol. 28, no. 15, pp. 4038–4052, 2016.

42. Z. Xu, X. Wei, Y. Liu, L. Mei, C. Hu, K.-K. R. Choo, Y. Zhu, and V. Sugumaran, "Building the Search Pattern of Web Users Using Conceptual Semantic Space Model," *International Journal of Web and Grid Services,* vol. 12, no. 3, pp. 328–347, 2016.

43. Z. Xu, X. Luo, Y. Liu, K.-K. R. Choo, V. Sugumaran, N. Yen, L. Mei, and C. Hu, "From Latency, through Outbreak, to Decline: Detecting Different States of Emergency Events Using Web Resources," *IEEE Transactions on Big Data,* vol. PP, no. 99, 2016.

44. Z. Xu, J. Xuan, Y. Liu, K.-K. R. Choo, L. Mei, and C. Hu, "Building Spatial Temporal Relation Graph of Concepts Pair using Web Repository," *Springer Information Systems Frontiers,* 2016. DOI: 10.1007 /s10796-016-9676-4.

7

PRICE-BASED DATA DELIVERY FRAMEWORK FOR DYNAMIC AND PERVASIVE IoT*

FADI AL-TURJMAN

7.1 Introduction

Internet of things (IoT) is a pervasive technology for applications ranging from smart grid to vehicular networking and smart homes to smart workplaces. IoT is growing as a framework to encompass all identifiable things in a dynamic and interacting network. The promise of clever approaches and dynamic systems that could benefit from the aggregation and analysis of information over the IoT infrastructure is quite pervasive. Scientists in networking, R&D divisions, and many businesses are in the race to develop an achievable and robust architecture to realize the IoT paradigm [1,2].

Yet, many hindrances render the IoT framework mostly a challenge. To date, much has been proposed on the promise and benefits of IoT, yet far less has covered the routing protocols to actually operate such a dynamic and large-scale paradigm [3,4]. The vision, however sparse, promises a robust and dynamic framework to integrate many enablers that are already outshined in research and development.

Obviously, wireless sensor networks (WSNs) are envisioned to play a dominant role in IoT frameworks. The resilience, autonomy, and energy-efficient traits of WSNs render them a vital candidate for dominating the information collection task of an IoT framework [5,6].

* This article was originally published in *Pervasive and Mobile Computing*. F. Al-Turjman, Price-based data delivery framework for dynamic and pervasive IoT, 2017. Reprinted with permission.

Equally vital, the use of RFID technologies for non-LOS and seamless identification of objects is gaining much prominence as a key player in IoT frameworks [7]. The low cost associated with deploying RFID tags (passive or active) is an important motivation. In fact, some argue that RFIDs have been a main motivator for the IoT framework [7,8].

The integration of these enablers, along with Internet-based and context-aware services, facilitate a dynamic platform for the IoT. Nevertheless, much of current research has focused on developing these enablers in segregation and optimizing their performance under local constraints and objectives. One of the most important tasks to be carried out, in such a large scale and dynamic environment, is relaying information from a source to a destination, given the new emerging characteristics in IoT. Typical routing approaches consider that all components belong to the same owner/provider, hence routing costs and link weights are directly proportional to their local provider characteristics. Though, IoT routing becomes inherently complex by multiple factors. An intrinsic design factor in IoT is delay-tolerance [9].

In reality, an IoT node has only partial knowledge regarding the full path to the destinations assigned to the packets it delivers. Due to splitting, which is mainly caused by nodal mobility, connectivity may occur on an irregular basis. In such circumstances, nodes are required to store and carry data packets until an appropriate forwarding chance ascends in a store–carry–forward fashion [10]. Typical sensor networks' routing approaches are unfortunately mobility-intolerant since most of the WSN network architectures assume stationary sensor nodes [11,12]. As stated earlier, we adopt an expanded notion of sensor networks that incorporates MANET nodes. An abundance of routing-layer protocols have been proposed to accommodate the dynamic topology in MANETs and WSNs [11,12].

Yet for all these protocols, it is implicitly assumed that the network is connected and there is a contemporaneous end-to-end path between any source/destination pair. In other words, the topology in the standard dynamic routing problem is assumed to be always connected and the objective of the routing algorithm, hence, is confined to finding the best currently available full path to move traffic from one end to the other. Unfortunately, none of these assumptions stand in a delay-tolerant setup. An IoT data delivery scheme must be delay-tolerant to cope with intermittent connectivity, in addition to providing faster delivery alternatives for other delay-sensitive types of data that demand minimal delays.

Furthermore, most entities participating in sensing, identifying, and relaying in IoT belong to different networks with multiple owners. It is not in the best interest of such networks to allow its resources to be utilized for relaying data across the network without compensation. For example, an intermediate relay node, belonging to a WSN for surveillance, would not freely take part in relaying information of nearby RFID readers or other WSNs. Thus, price and trading, in addition to all of the routing metrics that govern a mesh ad-hoc network, need to be considered before a suitable routing protocols is presented to relay packets across an inherently diverse IoT.

To this end, we define an *IoT setting* by the following four main characteristics: (1) Cost-effectiveness, (2) seamless integration, (3) reliability and trust, and (4) delay-tolerance. Hence, we provide a framework, encompassing a cost-efficient IoT architecture, to address data delivery objectives according to the aforementioned characteristics of the IoT setting. The design objectives are to be met with respect to metrics such as delay-tolerance, cost, and power-saving. Our proposed framework makes use of ubiquitous relays available in today's topologies to enhance connectivity and delivery rates between the components of the integrated topology. Our framework will as well provide delivery guarantees with respect to delay and connectivity over end-to-end links. Such guarantees will be carried out by dedicated components of our integrated architecture, in addition to other components incorporated within the wider IoT vision.

Our impact in this work comes in twofold. First, presenting a routing approach customized for the heterogeneous IoT components. This is only possible with our second contribution, a pricing model which caters for the diverse requirements and conditions of nodes which are willing to relay IoT data packets without using the Internet backbone. The pricing model presented here joins measures of load balancing, delay, buffer space, and link maintenance. An outline of the targeted routing problem and the dynamic constituents of the envisioned IoT is depicted in Figure 7.1.

The rest of this chapter is organized as follows. Section 7.2 covers the background on IoT routing and its enabling technologies. Then a rigorous definition of our proposed network model, manifesting the interactions of components in the IoT, and their governing constraints is proposed in Section 7.3. Next, we formally present our adaptive routing protocol in

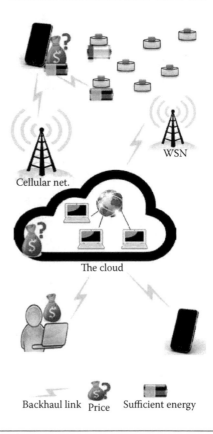

Backhaul link Price   Sufficient energy

Figure 7.1   An outline of cross-network routing in the IoT and the pricing forced by the network heterogeneity.

Section 7.4. Our proposed model is verified in Section 7.5 via use-cases and Markov-chain in Q-theory. Extensive results are performed and described in Section 7.6. Finally, our work is concluded in Section 7.7.

7.2 Background

Nowadays, everywhere around us is surrounded with different types of networks. WiFi, LTE wireless communications, broadcasting, streaming, and so on are quite common widely spread technologies. However, they bring their own limitations. These limitations can be in the form of cost or technology. Most often, it is about the cost of maintaining and placing an efficient network that can integrate all for what we call the IoT. Several attempts have been made for improvements and performance gains in the enablers of the IoT (especially WSNs and RFIDs). To present a perspective on these enablers, and the major domains of properties, Table 7.1 summarizes three main paradigms to the IoT. Accordingly, we emphasize two major driving forces. First, the lack of a distinctive routing approach that caters for dynamic IoT. The second drive lies in the tradeoff costs of routing over multiple entities, belonging to different service providers.

A major misconception was imposed by an inherent property of the IoT; namely, being a descendent of the Internet. That is, as research on the IoT developed, it was expected that a significant pool of protocols previously developed for Internet services would migrate into the IoT. Nonetheless, as the IoT is set to encompass many stationary (static WSNs, RFID readers, etc.) and dynamic (laptops, PDAs, cell phones, etc.) components, we are challenged with multiple issues [2,13]. Most importantly, assuming that all components will intercommunicate via the Internet is insufficient and often degrades the intended dynamic paradigm performance.

A major obstacle would stem from the mounting number of messages that overload a network already handling millions of hosts. This is a noteworthy problem as recent endeavors are targeting higher levels of dynamic interaction between the IoT and its users, as in the human–computer interaction work presented by Kranz et al. in [14]. As such, if a WSN needs to identify an object, with the aid of an RFID reader, direct communication between a sensing node (SN) and the reader would influence bottlenecks of communication and swarming the backhaul over the

Table 7.1   IoT Enablers and Their Properties

PROPERTY		WIRELESS NETWORKS		
	IoT	MANETS	WSNS	RFIDS
Topology	Dynamic	Dynamic[b]	Mostly static	Application dependent
Buffer size	Varies	High	Low	None
Medium contention	High	High	Medium	Low with singulation
Mobility	Frequent	Varies[b]	Limited	Frequent
Communication range	Varies	High	Medium (varies)	Reader dependent
Typical density	Very high	Small to medium	Medium to high	Medium to high
Computational power per node	Varies	High	Low	Low to none[a]
Internode communication	Heterogeneous		Homogeneous	

[a] Disregarding active-tags, as they equate many features of sensing nodes.
[b] Since Manets encompass VANets as well.

Internet with numerous packets. This is a prominent architecture, one that is strongly pushed for as a truly integrating IoT [15]. There is a need for establishing a cooperative scheme for routing in the IoT; one which includes all nodes with capabilities of relaying data. This includes those with only one access medium (e.g., WiFi routers) and others with multiple mediums (e.g., cell phones). Yet, due to obvious reasons of resource conservation, such entities would not participate in relaying data packets unless there is an incentive [16]. It is vital to note that some components only generate data (e.g., IDs), such as RFID tags.

Different incentives take part in the pricing model that dictates the choice of a group of candidates for relaying. Recent results in incentive based routing have been well studied. Zhong et al. present an elaborate study on routing and forwarding in MANets by emphasizing a scheme that ensures optimal gain for the individual nodes [17]. Auction pricing patterns [18] allocate resources to users through a bidding process conducted by the users. Auction pricing can accomplish equally resource allocation and service attributes. The scheme is based on profile bids where the seller computes an allocation to be given to the buyer. This sequence is repeated until all parts agree. We note that the lengthy negotiation process is ineffective particularly for mobile users while in high speed transit.

Dynamic priority pricing schemes [19], on the other hand, are applied on a wireless link shared by the subscribers, divided into different priority classes by the service provider. The mobile subscriber is allowed to select the preferable transmission rate, and its traffic allocation. In addition, the subscribers can split their traffic among several priority classes, and be charged accordingly. The efficiency of this scheme is in its simplicity and its scalability. The provider's profit increases according to the user's satisfaction by the service. Priority schemes assume, however, that the network's capacity buffer is not exceeded and priority thresholds are kept under a maximum level. Other schemes have been presented to incorporate dynamic game theory models, for noncooperative scenarios where local utility functions dictate the participation of nodes in relaying [20,21]. It is important to note as well that many of such factors are nontrivial to compute, and many nodes in the IoT would not possess the computational capacity to compute and execute local utility functions. Thus, it is intuitive to pursue a game theoretic approach for the IoT only if it caters for offloading the task of computing local utility functions to nearby high-end nodes.

Other problems stem from scalability issues in IoT, being an architecture that is envisioned to span continents and the globe [22]. The major issue is being able to maintain end-to-end links, and keeping track of nodes that are dynamically entering and exiting from the network. Remedies have been proposed by increasing the density of backhaul connections and multiple readers to enhance connectivity and capacity, respectively. However, recent studies highlighted the degrading effect of interreader and relay collisions. Ali et al. introduced a redundant reader elimination scheme to optimize tag coverage yet limit reader-to-reader contention; in addition to reducing the costs of deployment [23].

7.3 IoT System Model

Many factors are intrinsically dominant in the operation of a routing protocol. More factors are further augmented as we devise a routing protocol for the IoT paradigm with dynamic topologies and heterogeneous data generating/sharing systems in place. In such a comprehensive paradigm, an incentive data-sharing policy is required to motivate sensor owners to participate in the sensing process and to ensure that the provided data is fairly priced. And this in turn necessitates addressing IoT-specific challenges, such as system's limitations in terms of lifetime, available capacity, reachability, and delay. In addition, a careful focus on quality management and assurance constraints is to be considered, as well.

Thus, it is the scope of this section to detail and elaborate upon the factors that are considered in IoT-specific routing protocols that tackles all the above mentioned concerns. No single protocol would achieve all objectives, as many objectives are inherently contradictory, thus routing belongs to the notorious NFL (no free lunch) class of algorithms.

Our system is presented in the remainder of this section and elaborated upon in four components. First we present the IoT network as a whole, elaborating on the description of heterogeneous nodes in this model. Each of the resources pertaining to these nodes, and affecting the relaying scheme, are discussed in the following subsection. The discussion is completed with a derivation for the utility functions that would govern the choice of nodes, and finally the types of messages exchanged in the routing scheme to optimize upon the resources and residual energy in these nodes. In Table 7.2, a summary of the used notations is presented.

Table 7.2　Summary of Notations

NOTATION	DESCRIPTION
N	Number of in-network devices.
n_i	A node/device $i \in N$.
δ	A threshold on number of hops per routed packet.
Ψ_i	A quintuple computed for each $n_i \in N$ based on residual energy, delay, trust, and capacity per buffer.
u_i	Available storage capacity to compute and relay a message at n_i.
u_i'	Normalized buffer capacity per n_i.
π_i	Power consumption per n_i.
π_i'	Normalized power consumption per n_i.
E_i	The maximum energy budget per node n_i. It varies from one node to another in heterogeneous IoT network.
D_k	K^{th} data packet size in the queue.
E_{ij}	Euclidian distance between a source node i and a destination j.
ω	A delay step; which is the distance a wireless signal would travel in one-time unit.
D_{single}	A single hop delay a packet will experience.
D_{total}	The total end-to-end delay a packet will experience.
T_{D_j}	A normalized value representing trust level of the exchanged packets between a node j and the destination D.
P_r	The probability to be connected within r communication range.
γ	Path loss exponent in a specific environment.
μ	A normally distributed random variable with zero mean and variance σ^2.
K_0	A constant value calculated based on the mean heights of the transmitter and receiver.
λ_d, λ_t	The arrival rates for data and trusted packets, respectively.
μ_d, μ_t	The departure rates for data and trusted packets, respectively.
μ_{cd}	The rate of departures caused by finding better price in the system.
MQL	Mean queue length in the system.
RT	Response time in of the system.
P_{ij}	The probability of being in an ij state shown in Figure 7.3b.
γ	The average percentage of transmitted packets that succeed in reaching the destination.
β	The inflection point of the randomly generated packets sequences.
ϵ	Represents the tolerance to variation in data quality expressed in a Sigmoid function according to Equation 7.5.
α	A constant that determines the rate of decrease of the utility function in Equation 7.5.

7.3.1 IoT Model

We assume a network of heterogeneous devices, those belonging to WSNs, MANets, RFIDs, and stationary/mobile devices. Each communicating entity of these devices (i.e., wired/wirelessly enabled device) is considered as an active node in this design; hereon referred to as a *node*. Thus, given a set N covering all these devices, we represent each node as $n_i \in N$ where $i = \{1, 2, \cdots, |N|\}$. Thus the set N includes both nodes that are sole relays (access points, routers, WSN sinks, etc.) and other devices with relaying capabilities (communication and processing). We assume each n_i is connected to the network, as disconnected nodes would not take part in this scheme. That is, if there's no link from a node n_j to some other node $n_i \in N$ then $n_j \notin N$. It is important to note that the size of N varies over time as nodes enter, leave, and run out of energy.

Connectivity between nodes is assumed to take one of two modes. If nodes are in close proximity, then we advocate for direct communication between the nodes without rerouting through the Internet (via a backhaul). However, to sustain the important large scale aspect of the envisioned IoT, we dictate that packets traveling over a threshold of hops δ would be routed through a backhaul as an intermediate stage, and then rerouted to the final destination from the closest backhaul to that destination. It is thus an important factor to cater for both short and long range communication between nodes, both directly or via the Internet backbone. We remark again the importance of obeying to approaches that utilize the Internet backbone only when necessary, and reroute spatially correlated data packets between neighboring nodes without loading the backbone.

7.3.2 IoT Node

Each node $n_i \in N$ takes part in relaying, as well as other tasks. Accordingly, each n_i encompasses a group of resources, with a minimum of communication and processing units. Moreover, in the case of cell phones, PDAs, WSN sinks, and RFID readers, they would all encompass a larger pool of resources, not necessarily geared toward the routing task. Thus it is important to consider how the load of performing these tasks could affect/hinder the relaying capabilities of such nodes. We note their existence but in this scope we account for their effect on residual energy and buffer capacity. And thus, main design aspects considered in the utility function of an IoT-specific node is depicted in Figure 7.2(a). A quintuple Ψ_i is computed for each $n_i \in N$ aggregating the following parameters, for both their direct and implied effect on the routing scheme:

7.3.2.1 Residual Energy and Power Model Each node operating on battery power would possess an energy reservoir, denoted by e_i where $0 \leq e_i \leq E_i$. Here we denote E_i as the maximum charge for n_i, since this varies across the different types of nodes.

To normalize this representation across the heterogeneous nodes in this protocol, we define

$$\acute{e}_i = \frac{e_i}{E_i} \tag{7.1}$$

Knowing the size of data packet $\mathbf{D_k}$ to be forwarded, its distance to its next hop and the current load (u_i), each node would compute a value for the power consumption to be incurred by processing a given packet. The power consumption would be represented as π_i. However, since this is a crude number dependent on the available resources at node n_i and their strength (of transceiver), this value is normalized by dividing by its maximal attainable load and transmission distance. This would favor high end nodes with longer transmission capabilities and more buffers. The normalized value is represented as $\acute{\pi}_i$.

7.3.2.2 Load and Buffer Space Since an intermediate node might be taking part in multiple tasks, each node will represent its available capacity to compute and relay a message as a utilization factor u_i, which will be normalized by opposing it to its maximal capacity, thus yielding a normalized $\acute{u}_i$. This is directly derived from memory and processing operations, and the yield of the node's MCU in handling different paths. Our delivery approach adopts a data delivery approach where a network intermediate node i has a limited capacity for the maximum amount of data that can be relayed over a specific time period. We define a normalized relaying capacity for the set of i's as

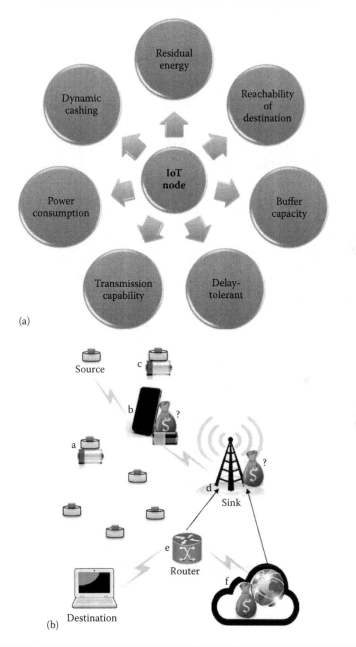

Figure 7.2 (a) The IoT-specific aspects incorporated in the pricing model for computing a utility function. (b) Use case demonstrating the routs taken by INs from the source to a destination (e.g., remote laptop).

$$\acute{u}_i = \frac{u_i}{max\_u_i} \tag{7.2}$$

where max_u_i is the maximum expected capacity.

7.3.2.3 *Delay*   We define a delay step ω which is the distance a wireless signal would travel in one-time unit. Let E_{ij} be the Euclidian distance between a source node i and a destination node j, then the discrete propagation delay over a single-hop link (i, j) would be $\frac{E_{ij}}{\omega}$. Hence, the discrete delay over a multi-hop path is the sum of the discrete delays of single-hop links that constitute that path. For the sake of generality, single-hop-delay (D_{single}) and total-delay (D_{total}), can be defined in Equation 7.3 and Equation 7.4, as follows:

$$D_{single} = \frac{E_{ij}}{\omega} + \psi \tag{7.3}$$

and

$$D_{total} = \sum_{total\ hops} D_{single} \tag{7.4}$$

7.3.2.4 *Trust*   Trust parameter is a history-based function that is calculated at the network intermediate node per destination to represent a D_j fulfillment measure. A higher T_{D_j} indicates that previous data exchanges between *node_i* and D_j have been fulfilled according to the predefined IoT characteristics (e.g., capacity, delay, trust) promised by D_j. We remark also that this delay parameter can be defined alternatively according to the applied IoT application with varying weighting factors. For example, delay is a key parameter in risk management IoT applications. However, it can be more relaxed in other kind of entertainment applications.

7.3.3 *Pricing Model*

All the previous factors are pertaining to nodal resources and their operation levels, in contrast to the remaining energy each node could support. However, an important aspect to cater for, and possibly arbitrate upon, is the price the nodes are going to charge for relaying a given data packet. That is, since the heterogeneous nodes in the IoT system do not belong to the same network nor the same owner, it is imperative that a monetary cost would be associated with the forwarding action [24,25].

This is an important aspect for integrating multiple heterogeneous nodes in the architecture and enhancing global scalability. The argument for utilizing current resources with a given cost/price is more dominant than claims of deploying enough resources to cater for all connectivity and coverage tasks of the envisioned IoT. We hereby adopt and build upon the former argument an IoT-specific pricing model.

Pricing schemes in heterogeneous networks such as the ones in IoT paradigm usually cover a wide range of factors to determine the value of a resource in usage. However, the most efficient schemes capitalize on the differential values of each of the heterogeneous components. We built our pricing strategy based on the original laws of supply and demand, the abundance of resources and their homogeneity decrease their value. And higher prices are usually assigned to nodes with rare services [26,27]. Moreover, an IoT-driven pricing model has to realize a level of service that aggregates data from several sources, including the network-context (e.g., 4G, WiFi, WiMax), the mobile apps' pool, and other sources, to produce better reliable readings. We remark also that an IoT-specific pricing model shall not assume a direct provider–client relation in determining their price mechanisms. In contrast, in large economic systems such as the one we are targeting in this research, entities known as intermediate nodes (INs) are required for coordinating network management tasks. These INs take care of the necessary authentication, billing, and interfacing tasks to find the appropriate service provider within the heterogeneous crowd of resources in the IoT market. To this end, we define an IoT-specific setting by the following four main characteristics: (1) Node residual energy, (2) load and buffer space, (3) trust-level, and (4) delay-tolerance. Those characteristics have been adopted into two different simulation environments; MATLAB® and Simulink® in order to validate our price-based results. Simulink®, a framework built on MATLAB®, is used for validation purposes in order to obtain more realistic results by imitating the real multilayered net-working process.

Hence, we provide a pricing framework for each node, encompassing a cost-efficient IoT architecture, to address data delivery and routing objectives according to the aforementioned characteristics of the IoT setting. We introduce γ_i, which is a pricing factor for each node in the IoT. This is a factor that could be set as a flat rate per number of bytes transmitted or computed based on the state of the current resources at node n_i represented by Ψ_i. In this work we adopt the latter, as a proof of concept to the monetary exchange for forwarding in the IoT under varying conditions. Thus, we denote the price charged by each node n_i as p_i:

$$p_i = \gamma_i * \left[\frac{E_{Tx}(D_k, n_j) + E_{Rx}(D_k)}{e_i} + \acute{\pi}_i + \acute{u}_i \right] \qquad (7.5)$$

It is intuitive to note that owners of nodes in the vicinity of such a network, may choose to adaptively contribute or withdraw from the topology by varying the value assigned to γ_i. That is, setting it to a relatively high value would diminish the chances of it being selected for relaying.

7.3.4 Communication Model

In practice, the signal level at distance d from a transmitter varies depending on the surrounding environment. These variations are captured through the so called log–normal shadowing model. According to this model, the signal level at distance d from a transmitter follows a log–normal distribution centered on the average power value at that point [28]. Mathematically, this can be written as

$$P_r = K_0 - 10\gamma \log(d) - \mu d \qquad (7.6)$$

where d is the Euclidian distance between the transmitter and receiver, γ is the path loss exponent calculated based on experimental data, μ is a normally distributed random variable with zero mean and variance

σ^2, that is, $\mu \sim \mathcal{N}(0, \sigma^2)$, and K_0 is a constant calculated based on the mean heights of the transmitter and receiver.

7.4 ARA Routing Approach

The integrated architecture imposed by the heterogeneity of the IoT demands a scalable and inclusive routing protocol. The latter property refers to the exploitation of different relaying resources that are able to carry forward a data packet towards the destination. This section presents ARA protocol.

ARA is divided into two stages: Forward and backward. The forward stage starts at the source node by broadcasting setup messages to its neighbors. A setup message includes the cost seen from the source to the current (intermediate/destination) node. A node that receives a setup message will forward it in the same manner to its neighbors after updating the cost based on the values computed in Ψ_i. All setup messages are assumed to contain a route record that includes all nodes' IDs used in establishing the path fragment from the source node to the current intermediate node. The destination collects arriving setup messages within a route-select (RS) period, which is a predefined user parameter.

The backward stage starts when an acknowledgment (Ack) message is sent backward to the source along the best selected path (called *active* path) in terms of the parameters passed in Ψ_i. If a link on the selected path breaks (due to node movement or bad channel quality), the Ack at an intermediate node i is changed to setup message (called i_setup) and forwarded to neighbors of i which has discovered the error.

Once the source receives the i_setup, the active path between S and D is established. When no breaks are discovered, the source receives an Ack and knows that the path has been established, and it starts transmission. If during the communication session (i.e., after selecting the active path) a break is detected, the intermediate node detecting the break will send data on an alternative route (if any) or it will buffer data and send an i_setup message to the destination to look for an alternative path.

In general, nodes can learn about their neighbors and update the routing table (RT) either by receiving a broadcasted setup message and accordingly updating its neighborhood table, or by broadcasting a "hello" message periodically, if no messages have been exchanged. This "hello" message is sent only to the neighborhood of the node. A new neighbor, or failing to receive from a node for two consecutive "hello" periods, is an indication that the local connectivity has changed.

A pseudocode description of the source node algorithm is shown below. Lines 1–2 represent the beginning of the forward stage, where a request to establish an active path is initiated. Such that, if S has new packets to send and no route is known to targeted destination D, then a setup message is forwarded to all available neighbors of S. To do so, all INs node broadcast their identity at the deployment stage and each S node keeps a record of the next hop towards some IN. Each source node n_i has a next node record that has the following: *ID field* to recognize the next relaying intermediate node ID; *Geo_Loc* field to determine node geographical coordinates; and *Number_of_hops field*, which has the number of hops towards the destination D. Note that this process will construct a price-based tree for each S node, such that the tree of INs that is rooted at S and involves all price-efficient INs toward the destination D will be identified at the initialization of the network. Lines 3–4 indicate that the path has been found.

Algorithm 7.1 for Source Node S

1. **If** S has a new *data* msg & no route to D
2. **Then** forward a *setup* msg.
3. **If** S receives *D_Ack* or *i_setup* msg,
4. **Then** check local p_i and send the new *data* msg's if satisfied.
5. **If** S doesn't receive a response for a RD period,
6. **Then** go to line 2.
7. **If** no pkts are exchanged for *hello_interval* time units,
8. **Then** send a *hello* msg and update RT and p_i.

Hence, active path between S and D is updated and source begins transmitting the new data packets. Lines 5–6 describe the case where a route discovery (RD) period has expired. Therefore, the source restarts the route discovery process by sending a new setup message. Finally, lines 7–8 indicate that S has not exchanged messages with neighbors for more than *hello_interval* time units. Thus a "hello" message is sent and RT is updated accordingly.

A pseudocode description of the intermediate node algorithm is shown below. Lines 1–2 handle the forward stage, such that if an intermediate node i receives a setup message, it forwards this message to all its unvisited neighbors and records every visited node to establish a backward path. Contrarily, lines 3–7 handle the backward stage of the algorithm. If node i receives Ack from destination (called D_Ack), then it checks whether the neighbor toward S on the backward path is reachable or not (i.e., has a broken link). If reachable, it passes the D_Ack to this neighbor and records the necessary information to establish the active path. Otherwise, it initiates a new setup process between i and S, by sending i_setup message to i's neighbors. Lines 8–9 keep forwarding this i_setup message until it reaches S to establish an active path between i and S instead of the broken one.

Similarly, lines 10–12 check for the availability of the next hop on the active path while data packets are transmitted through i toward the destination D. If next hop is not available, the intermediate node i checks for an alternative path. If a new path has been established, lines 13–14 detour the data packets between S and D along this new partial route and update the active path. If no alternative path is found, line 15 buffers the data packets and initiate a new setup process. We remark that lines 2 and 9 will kill any setup message, if i is not willing to participate in routing.

Algorithm 7.2 for Intermediate Node i

1. **If** i receives *setup* msg,
2. **Then** check thresholds and update/forward *setup* msg if satisfied. Also, the forwarded *setup* msg records visited nodes while traveling to D.
3. **If** i receives *D_Ack*
4. **Then, If** a backward_neighbor is reachable,
5. **Then** forward the *D_Ack*
6. **If** backward_neighbor is not reachable,

7. **Then** send an *i_setup* msg and update RT and local p_i.
8. **If** *i* receives *i_setup* msg
9. **Then** check thresholds and forward *i_setup* msg if satisfied. Also, the forwarded i_*setup* msg records visited nodes while traveling to destination.
10. **If** *i* receives *data* msg
11. **If** next hop is still reachable
12. **Then** send *data*
13. **If** a new active path was established
14. **Then** check the price, update RT and send *data* if satisfied.
15. Else buffer data and send i_setup

Finally, a pseudocode describing the algorithm at the destination node D is shown below. Lines 1–10 handle the case when a setup process has been initiated by an intermediate node *i*. This also indicates link breakage at node *i* in active path between S and D. If there exists alternative path(s) passing through the node detecting link breakage (i.e., node *i*) or passing through the source S, lines 3–4 select the best-cost path and notify *i*. Otherwise, lines 5–10 initiate a new setup process and act as a source node in looking for a new path to S. Therefore, it sends to all D's neighbors and waits for an Ack from the source S (called S_Ack). Meanwhile, lines 11–14 represent the backward stage in response to the forward stage that has been initiated at S. The destination D keeps receiving setup messages with the corresponding found paths between S and D for a route select (RS) interval. After RS time units, D acknowledges the source S that the active path has been established by sending a D_Ack message to it through the best-cost selected path.

Algorithm 7.3 for Destination Node D

1. **If** *D* receives *i_setup*
2. **Then** remove paths containing broken links.
3. **If** there exist path(s) passing through *i* or S
4. **Then** select best-cost path and notify *i*.
5. **If** no paths found
6. **Then** send a *setup* msg
7. **If** *D* receives *S_Ack* or *i_setup*

8. **Then** select path indicated by received msg.

9. **If** *D* doesn't receive a response for a RD period,

10. **Then** go to line 5.

11. **If** *D* receives *setup* msg RS not expired

12. **Then** store the candidate path and cost.

13. **If** RS expired

14. **Then** select best-cost path and send *D_Ack* on it.

7.5 Use Case and Theoretical Analysis

To demonstrate the utility of the ARA protocol, we hereby adopt a use case that utilizes heterogeneous nodes in a sample IoT environment. The remainder of this use case will refer to Figure 7.2. A sensing node (the source) has obtained information to be sent to a destination computer. However, no direct link connects both devices, and intermediate devices belong to different networks.

We assume that nodes *a*, *b*, *c*, *d*, *e*, and *f* are all willing to relay, yet *a* and *c* are already depleted in energy. The sink, node *d*, is powered by electricity and acts as an intermediate node between the resourceful cell phone *b* and the router *e*.

ARA will initiate a setup message sent to *a*, *b*, *c*, and its current neighbors. Since *a* and *c* have depleted batteries, they will terminate the flow of the setup request toward the destination. Since the cell phone *b* is in range of communication to the source, it will forward the message to its neighbors (not highlighted here as the pattern is clear).

Eventually the shortest path to the destination is established. The destination will receive two streams $\{S \rightarrow b \rightarrow d \rightarrow e \rightarrow D\}$ and $\{S \rightarrow b \rightarrow d \rightarrow f \rightarrow e \rightarrow D\}$. Since both f and e are resourceful entities, the arbitration of number of hops would manifest a preference for the former route, which will carry an Ack message back to the source node.

It is important to note that an Internet link (both forward and backward), which would also incur a cost, takes part in the route options, as the setup message would also parse through it when it is beyond the preset threshold of hops dictated by the application and source request. Furthermore, the IoT network under study can be modeled using queueing theory for steady state evaluation with an abstraction as illustrated in Figure 7.3(a).

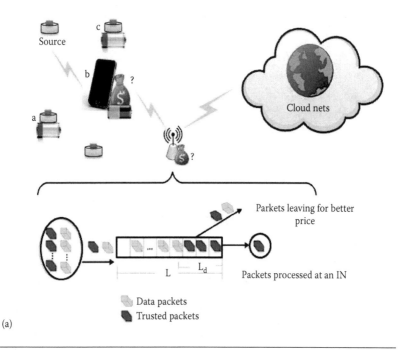

(a)

Figure 7.3   (a) The Q-model for the IoT networks.   (*Continued*)

The resulting continuous time Markov chain would be a multi-dimensional one for multiple types of traffic similar to the one presented in Figure 7.3(b). Where λ_d, λ_t are the arrival rates and μ_d, μ_t are departure rates due to service completion for data and traffic packets, respectively. μ_{cd} is the rate for departures caused by finding better price instead of service completion in the system. Various solution methods can be employed to solve such a system for steady state probabilities. Once the steady state probabilities are obtained, they can be employed for computation of QoS measures such as, mean queue length (MQL), throughput (γ), and response time (RT) as follows:

$$MQL_t = \sum_{i=0}^{L_t} \sum_{j=0}^{L} iP_{ij} \qquad (7.7)$$

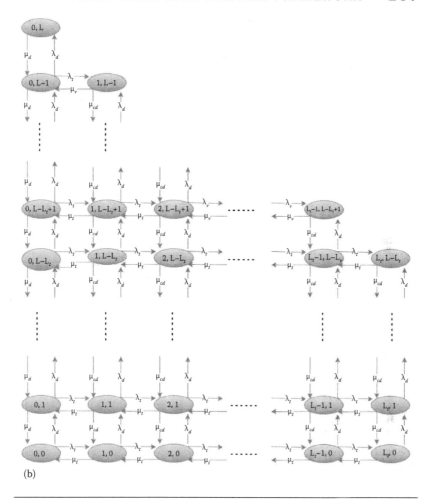

(b)

Figure 7.3 (Continued)   (b) The multi-dimensional Markov-chain for the IoT networks.

$$MQL_d = \sum_{i=0}^{L_d} \sum_{j=0}^{L} jP_{ij} \qquad (7.8)$$

$$\gamma_t = \sum_{i=0}^{L_t} \sum_{j=0}^{L} \mu_t P_{ij} \qquad (7.9)$$

$$\gamma_d = \sum_{i=0}^{L_d}\sum_{j=0}^{L}\mu_d P_{ij} \tag{7.10}$$

$$RT_t = \frac{MQL_t}{\gamma_t} \tag{7.11}$$

$$RT_d = \frac{MQL_d}{\gamma_d} \tag{7.12}$$

Please note that it is also possible to employ similarly the steady state probabilities for computation of expected value for energy consumption of the system considered.

7.6 Performance Evaluation

In this section, the effectiveness of ARA is validated while assuming a set of in-network heterogeneous nodes. Simulation results show the performance efficiency in terms of average delay, price, idle time, and throughput in comparison to key approaches in the literature. In addition, the quality of the data delivery approach is assessed under varying rates per number of bytes transmitted, average energy consumption, and several counts of the network nodes.

7.6.1 Simulation Setup and Baseline Approaches

Using MATLAB® R2016a and Simulink® 8.7, we simulate randomly generated heterogeneous networks. The generated networks are random in terms of their nodes' positions and densities. In order to route data in these randomly generated networks, we apply our ARA scheme. The output of the ARA scheme is compared to output of another four baseline approaches in the literature. These baseline approaches address the same problem tackled in this research; however, they use different routing strategies. The first approach forms a minimum spanning tree (MST) to find the most reliable route in a heterogeneous sensor network

[29], and we call it the *MST approach*; the second is for solving a Steiner tree problem with minimum number of Steiner points [30], and we call it *Steiner tree* (ST); the third is for adaptive data delivery, which we call the *dynamic routing approach* (DRA) [31], and fourth one is called LinGO [28]. In all these baselines we assume a packet size is equal to 512 bits, which is a typical size for the IoT communication protocols. Every S node has an initial energy of 50 joules and generates 150 packets/round. A *round* is defined as the time span per which all S nodes have reported/ requested a piece of data.

The MST opts to establish an MST through selected multi-hop paths. It first computes an MST for the given source and destination nodes and then forwards messages over the minimum tree model in which it finds the least count of hops to maintain the best path cost. ST first combines nodes that can directly reach each other into one connected graph. The algorithm then identifies for every three connected graphs a node x that is at most r (m) away. Then these three connected graphs are merged into one. These steps are repeated until no such x could be identified (i.e., no isolated nodes). DRA takes into consideration the nodes' coordinates in order to limit the updates sent out by the any moving node to a local area. LinGO, which is a Link quality and Geographical beaconless OR protocol, introduces a different progress calculation approach compared to the aforementioned ones. It takes into account both the progress of a given forwarding node towards the destination with respect to the last-hop, as well as the radio range. In this way, LinGO reduces the number of required hops. Both MST and ST routing strategies are used as a benchmark in this research due to their efficiency in finding the nearest next hop towards the destination while maintaining the minimum number of required nodes in the source–destination path. On the other hand, DRA is chosen due to its efficiency in adapting to any newly generated topology due to node mobility/ heterogeneity. We remark that the original MST, ST, and DRA approaches are not hierarchical. Thus, we employed the modified versions of them to make them suitable for our proposed hierarchical framework, where the modified versions take into consideration the in-network nodes' heterogeneity and choose the next hop based on types of the surrounding node types. For example, a Zigbee-based IoT node will scan for another Zigbee-based node to be considered a candidate neighboring node for packet relaying/forwarding, since a larger network size implies longer paths, and thus, higher probabilities for heterogeneity. We examined the four data delivery schemes while the size of the network increases in terms of the IoT-nodes' count. Knowing that larger node count in a data path raises the risk of node failure and hence, dropped packets. Thus, choosing shorter peripheral paths is better for the overall quality/price gain.

The routing schemes DRA, MST, ST, and ARA are executed on 600 randomly generated wireless heterogeneous network topologies in order to get statistically stable results. The average results hold confidence intervals of no more than 2 percent of the average values at a 95-percent confidence level. We assume a predefined fixed time schedule for traffic generation at these networks. Data packets are delivered by applying these three approaches.

Based on experimental measurements taken in a site of dense heterogeneous nodes [18], we set the communication model variables and other simulation parameters as shown in Table 7.3. We adopt the described signal propagation model in Section 7.3 where the utilized variables/parameters values, shown in Table 7.3, are set to be as follows: $\gamma = 4.8$, $\delta = 10$, Pr = −104 (dB), and μ to be a random variable that follows a log–normal distribution function with mean 0 and variance of δ^2.

Table 7.3 Parameters of the Simulated Networks

PARAMETER	VALUE
τ	70 percent
n_c	110
ψ	0.001 (msec)
D_{max}	500 (msec)
ω	200,000 (km/s)
γ	4.8
δ^2	10
P_r	−104(dB)
K_0	42.152
r	100 (m)

Moreover, we assume heterogeneous transceivers communication ranges to validate our results in a typical IoT setups. For validation and verification purposes, we also used MATLAB® with Simulink® Framework [32]. Simulink can support wireless channel temporal variations, node mobility, and node failures. The simulations last for two hours and run with the log–normal shadowing path loss model. In Simulink, we adopted also the same path loss and physical layer parameters shown in Table 7.3.

7.6.2 Performance Parameters and Metrics

To compare the performance of these three schemes, the following four performance metrics are used.

- Average delay: Is measured in milliseconds and is defined as the average amount of time required to deliver a data unit to the destination.
- Idle time: This metric reflects the ratio of idle time every node spend while just waiting to forward a message. It is measured in μ*sec*.
- Throughput: Is set here as a quality measure. It is the average percentage of transmitted data packets that succeed in reaching the destination reflecting the effect of node heterogeneity and delay in IoT setups over the utilized data delivery approach.
- Average price: This metric is used to observe the influence of the utilized data delivery approach on the overall price to deliver a data unit from source to destination on average.

Meanwhile, the three data delivery performance is assessed using the following three parameters:

- The size of the network in terms of total node count. This reflects the application's complexity and the scalability of the exploited routing scheme.
- Average energy consumption rate per data unit ($\bar{\pi}_i$) as an indicator of the network power saving.
- Cost (γ_i) to observe the influence of the charged price rate over the utilized data delivery approach.

7.6.3 Simulation Results

For a varying number of heterogeneous nodes (between 50 and 350) and deployment space (= 1200 Km2), Figure 7.4 compares ARA approach with DRA, MST, and ST in terms of data delivery latency. It shows how ARA outperforms the other approaches under varying network size. Unlike the other approaches, exchanged messages using the ARA do not show a rapid increment in the end-to-end delay while the network size is growing. This is because of the utilized utility function in Equation 7.5 that has been considered by ARA approach. Although DRA is the most adaptive approach, its delay is increasing rapidly while the network size is increasing. However, delay is slower in the monotonically increase while applying MST and ST approaches. It's worth mentioning also that when the network size is greater than or equal to 300, the ARA cannot improve any more in terms of delay. However, it's obvious that the ARA approach is achieving the lowest delay with respect to the varying network size.

Meanwhile, the overall network throughput levels achieved by the ARA outperform the levels achieved by other approaches due to considering the next hop status before forwarding the message to it, as depicted in Figure 7.5. In general, network throughput is increasing monotonically for all approaches while the network size is increasing. However, ST is the worst due to ignoring the current status of the node before message forwarding.

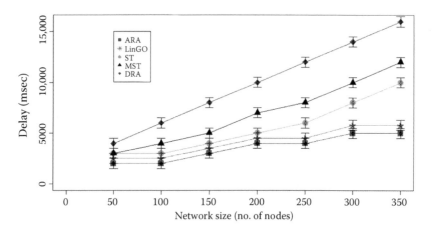

Figure 7.4 Latency vs. number of nodes.

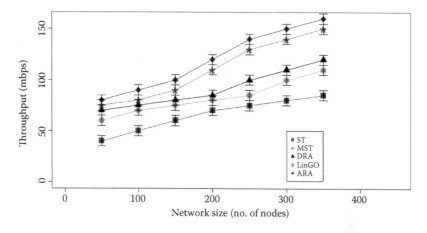

Figure 7.5   Throughput vs. the network size.

Furthermore, Figure 7.6 depicts the effects of the allowed average energy consumption level per node on the network throughput. It shows a monotonically increase in throughput for all approaches while varying the available energy budget. This comparison is performed while considering a fixed network size equal to 150 nodes, and average γ_i rate equal to 0.002 dollar/byte. Notably, more saving in terms of energy is achieved by applying the ARA and LinGO approaches. However, the ARA approach achieves the highest throughput with respect to energy. When the energy budget is greater than or equal to 60 (kilojoule), the network throughput is saturated due to other design factors such as γ_i and network size and capacity. Also, it worth

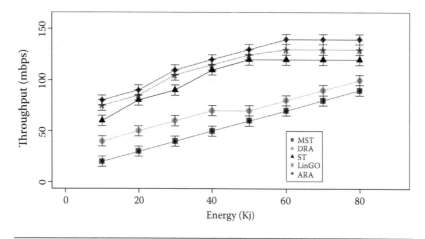

Figure 7.6   Throughput vs. the available energy.

remarking here that LinGO adds redundant packets in order to increase the packet delivery probability while experiencing link error periods. This leads to significant increments in the overall throughput.

In Figure 7.7, the network end-to-end delay is decreasing linearly for all methods while γ_i is less than or equal to 50 percent of the initial rate. This can be returned to the main objective of all these methods in providing the best QoS while considering the cost factor. However, ARA again has the best delay performance with respect to all other approaches due to direct influence of the utility function in Equation 7.5 in choosing the next hop toward destination. Also, it is worth noting that when the γ_i is greater than or equal to 50 percent, all approaches cannot improve any more in terms of the end-to-end delay. This has a great impact on the network QoS.

In Figure 7.8, average idle time is compared under varying total count of network nodes. All approaches are experiencing a monotonically increase in the average idle time while increasing the network size. This is expected due to the availability of several routing options/resources. In addition, we return the increase of idle time when network size is increased, to the dense distribution of network nodes within a fixed deployment space (= 1200 km²). Such a dense distribution provides idle resources as well. Nevertheless, ARA has the lowest average idle time in this comparison again due to the ability of the proposed utility function in Equation 7.5, where better resource management and utilization is guaranteed.

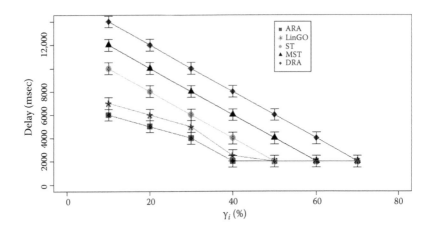

Figure 7.7 Delay vs. the average γ_i rate percentage.

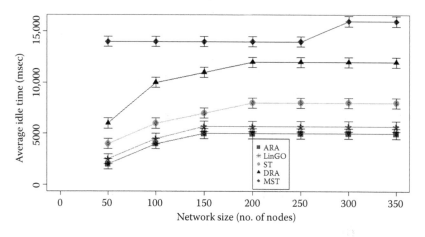

Figure 7.8 Average idle time vs. the network size.

In Figure 7.9, ARA consistently outperforms the MST, ST, LinGO, and DRA while increasing the nodes' counts. ST and DRA are the worst in terms of average price and that can be returned to their complexity in locating the next hop. Unlike ST and DRA, MST and LinGO approaches are very close to ARA. However, ARA still outperforms them. The reason is that LinGO and MST add redundant packets in order to increase the packet delivery probability while experiencing link error periods. This leads to significant increment in the overall price. In general, ARA is better because of the computed price factor based on the state of the current resources at every node n_i represented by Ψ_i.

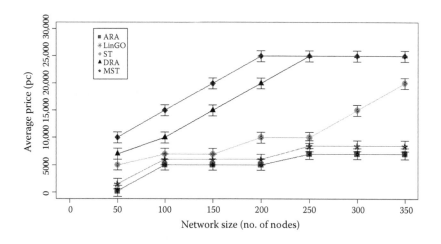

Figure 7.9 Average price vs. the network size.

It is noted also that the average price is increasing while the network size is increasing. Again, this has been accomplished due to considering longer routes while expanding the network size. We remark that as the nodes count increases, the total price achieved by ARA becomes more identical and minimal with respect to other approaches. This can be returned to the excess in the available nodes, and thus, the effect of better choices becomes observable and prevents any increment in the price. In general, the function in Equation 7.5 utilizes the aforementioned parameters after normalization, in a manner that maps the expected user experience to changes in individual utility parameters.

To show the impact of the aforementioned parameters on our utility function in Equation 7.5, we present the plots in Figures 7.10 through 7.12 for each of the utility parameters *Delay*, *Network Quality* (i.e., throughput), and *Trust T*, respectively. In Figure 7.10, *Delay* is plotted with a constant α that determines the rate of decrease of the exponential utility in Equation 7.5. This particular function was chosen for *Delay* to reflect the rate of loss in the quality of experience (QoE) [1] as delay increases. By varying the value of α, it is possible to achieve different levels of delay-tolerance as shown in Figure 7.10, where we chose α = 0.5 for delay-tolerant data and α = 0.1 for more delay-sensitive data. We note that, for a delay-sensitive data request,

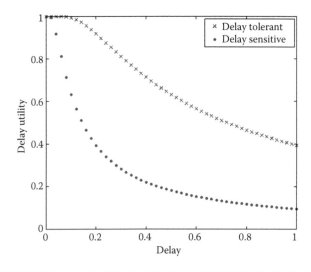

Figure 7.10   Delay function plot.

a very low delay has to be achieved in order to provide a high delay utility component.

The Quality parameter is plotted in Figure 7.11. We note that we adopt a Sigmoid function according to Equation 7.5 where the tolerance to variation in data quality is expressed by fixing the value of ϵ (set here to 10) and varying the inflection point denoted by the value of β. Thus, if the requested data is quality-sensitive (e.g., VoIP is sensitive to low transmission rate) the function will require a higher value before the utility increases (as depicted in the lower plot with $\beta = 0.8$ in Figure 7.11). In contrast, lower constraints on quality require a utility that increases rapidly at a lower value of *Quality*, which can be achieved with an early inflection point ($\beta = 0.5$ in Figure 7.11). The value of Quality in Figure 7.11 ranges from 0 to 1, where 1 indicates the best level of quality attainable depending on the quality metric.

Lastly, Figure 7.12 shows the plot of the Trust function where $T \in [0, 1]$. Note that an intermediate node can give more emphasis to this parameter through the factor σ to particularly penalize sinks with bad service accounts. This is shown in the lower plot of Figure 7.12 where $\sigma = 2$, whereas the upper (better T) plot is a result of $\sigma = 1$.

We note here that we multiply the Trust component by the rest of the utility function in Equation 7.5 in order to balance the effect of

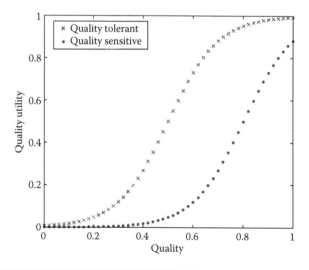

Figure 7.11   Quality function plot.

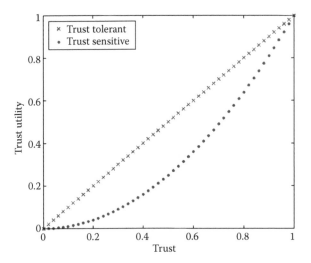

Figure 7.12   Trust function plot.

"deceiving" intermediate nodes that may offer attractively low P_i in order to pass false quality promises. Moreover, we divide the utility function by $P_{intermediate}$ in order to protect the client from situations where two or more intermediate nodes happen to achieve almost equal utility scores while charging prices that, although less than P_i, largely vary.

Evaluation results revealed that price-based routing scheme is better in most cases. Price-based routing schemes have longer path lengths when contrasted to shortest routing schemes, as expected. But they have had better results in terms of cost, latency, idle time, and scalability. ARA avoids using busy, centric nodes, thus data in ARA deliver over more nodes. Data reach destinations over available nodes instead of waiting for queue in busy nodes.

7.7 Conclusion

This chapter proposed a price-based routing scheme for heterogeneous IoT networks called the ARA. ARA aims at establishing a cap on the internodal routing price to dynamically utilize the Internet backbone if the source to destination distance surpasses a preset (case optimized) threshold. Promising simulation results are achieved by altering significant parameters, including network size and the available energy budgets, and seeing how it affects certain metrics in the IoT paradigm such as latency, cost, and resource utilization.

These results show the efficiency of our framework when compared to three prominent ad hoc data delivery protocols. Our simulation results show that the ARA exhibits superior performance for different network sizes, lifetime, end-to-end delays, quality, and prices. It is strongly recommended with huge size networks, as it is the most cost-effective one in the long run.

Future work would investigate utilizing IoT nodes in Smart City settings as mobile data collectors with semideterministic mobility trajectories. Also, of practical interest is the application of localization methods among sensors operating on different technologies and studying the effect of such methods on the system's performance and the delivery rate of the corresponding ARA scheme.

References

1. F. Al-Turjman, A. Alfagih, and H. Hassanein, "A Novel Cost-Effective Architecture and Deployment Strategy for Integrated RFID and WSN Systems," in *Proc. of the IEEE Int. Conf. on Computing, Networking and Communications (ICNC)*, Maui, Hawaii, pp. 835–839, 2012.
2. F. Al-Turjman, "Impact of User's Habits on Smartphones' Sensors: An Overview," *HONET-ICT International IEEE Symposium*, 2016.
3. L. Atzori, A. Iera, and G. Morabito, "The Internet of Things: A Survey of Computer Networks," *Computer Networks, Elsevier*, vol. 54, pp. 2787–2805, 2010.
4. G. Singh and F. Al-Turjman, "Learning Data Delivery Paths in QoI-Aware Information-Centric Sensor Networks," *IEEE Internet of Things Journal*, vol. 3, no. 4, pp. 572–580, 2016.
5. Q. Zhu, R. Wang, Q. Chen, Y. Liu, and W. Qin, "IOT Gateway: Bridging Wireless Sensor Networks into Internet of Things," in *Proc. of the IEEE/IFIP Int. Conf. on Embedded and Ubiquitous Computing (EUC)*, Hong Kong, China, pp. 347–352, 2010.
6. N. Aitsaadi, N. Achir, K. Boussetta, and G. Pujolle, "Artificial Potential Field Approach in WSN Deployment: Cost, QoM, Connectivity, and Lifetime," *Elsevier Computer Networks*, January 2011.
7. E. Welbourne, L. Battle, G. Cole, K. Gould, K. Rector, S. Raymer, M. Balazinska, and G. Borriello, "Building the Internet of Things Using RFID: The RFID Ecosystem Experience," *Internet Computing, IEEE*, vol. 13, pp. 48–55, 2009.

8. I. Fajjari, N. Aitsaadi, M. Pioro, and G. Pujolle, "New Virtual Network Static Embedding Strategy within the Cloud's Private Backbone Network," in *Elsevier Computer Networks*, April 2014.

9. L. Boloni and D. Turgut, "Should I Send Now or Send Later? A Decision-Theoretic Approach to Transmission Scheduling in Sensor Networks with Mobile Sinks," *Wiley's Wireless Communications and Mobile Computing Journal (WCMC)*, vol. 8, no. 3, pp. 385–403, 2008.

10. G. Wang, D. Turgut, L. Boloni, Y. Ji, and D. Marinescu, "Improving Routing Performance through m-Limited Forwarding in Power-Constrained Wireless Networks," *Journal of Parallel and Distributed Computing* (JPDC), vol. 68, no. 4, pp. 501–514, 2008.

11. G. Solmaz, M. I. Akbas, and D. Turgut, "A Mobility Model of Theme Park Visitors," *IEEE Transactions on Mobile Computing* (TMC), vol. 14, no. 12, pp. 2406–2418, 2015.

12. D. Turgut and L. Boloni, "Heuristic Approaches for Transmission Scheduling in Sensor Networks with Multiple Mobile Sinks," *The Computer Journal*, vol. 54, no. 3, pp. 332–344, 2011.

13. O. Vermesan, P. Friess, P. Guillemin, S. Gusmeroli, H. Sundmaeker, A. Bassi, I. Jubert, M. Mazura, M. Harrison, M. Eisenhauer, "Internet of Things Strategic Research Roadmap," *Internet of Things: Global Technological and Societal Trends*, 2009.

14. M. Kranz, P. Holleis, and A. Schmidt, "Embedded Interaction: Interacting with the Internet of Things," *Internet Computing, IEEE*, vol. 14, no. 12, pp. 46–53, 2010.

15. A. Sarma, and J. GirÃo, "Identities in the Future Internet of Things," *Wireless Personal Communications, Springer Netherlands*, vol. 49, pp. 353–363, 2009.

16. M. Afergan, "Using Repeated Games to Design Incentive-Based Routing Systems," in *Proc. of the IEEE Int. Conf. on Computer Communication* (INFOCOM), Barcelona, Spain, pp. 1–13, 2006.

17. S. Zhong, L. Li, Y. Liu, and Y. Yang, "On Designing Incentive-Compatible Routing and Forwarding Protocols in Wireless Ad-Hoc Networks," *Wireless Networks*, vol. 13, no. 9, pp. 799–816, 2007.

18. X. Wang and H. Schulzrinne, "Pricing Network Resources for Adaptive Applications," *IEEE/ACM Trans. Netw.*, vol. 14, no. 3, pp. 506–519, Jun. 2006.

19. G. V. Ozianyi, N. Ventura, and E. Golovins, "A Novel Pricing Approach to Support QoS in 3G Networks," *Computer Networks*, vol. 52, no. 7, pp. 1433–1450, May 2008.

20. W. Saad, Z. Han, M. Debbah, A. Hjorungnes, and T. Basar, "Coalitional Game Theory for Communication Networks," *Signal Processing Magazine, IEEE*, vol. 26, pp. 77–97, 2009.

21. G. Singh and F. Al-Turjman, "A Data Delivery Framework for Cognitive Information-Centric Sensor Networks in Smart Outdoor Monitoring," *Elsevier Computer Communications*, vol. 74, no. 1, pp. 38–51, 2016.

22. H. Sundmaeker, P. Guillemin, P. Friess, and S. Woelfflé, "Vision and Challenges for Realising the Internet of Things," *CERP-IoT, European Commission*, Luxembourg, 2010.
23. K. Ali, H. Hassanein, and W. Alsalih, "Using Neighbor and Tag Estimations for Redundant Reader Eliminations in RFID Networks," in *Proc. of the IEEE Int. Wireless Communications and Networking Conference* (WCNC), Quintana-Roo, Mexico, pp. 832–837, 2011.
24. F. Al-Turjman, "Cognition in Information-Centric Sensor Networks for IoT Applications: An Overview," *Springer Annals of Telecommunications*, pp. 1–16, 2016. DOI: 10.1007/s12243-016-0533-8.
25. F. Al-Turjman and H. Hassanein, "Towards Augmented Connectivity with Delay Constraints in WSN Federation," *International Journal of Ad Hoc and Ubiquitous Computing*, vol. 11, no. 2, pp. 97–108, 2012.
26. A. DaSilva, "Pricing for QoS-enabled Networks: A Survey," *IEEE Commun. Surveys & Tutorials*, vol. 3, no. 2, pp. 2–8, 2000.
27. D. Turgut and L. Bölöni, "IVE: Improving the Value of Information in Energy-Constrained Intruder Tracking Sensor Networks," in *Proc. of the IEEE ICC*, pp. 6360–6364, 2013.
28. D. Rosário, Z. Zhao, A. Santos, T. Braun, and E. Cerqueira, "A Beaconless Opportunistic Routing Based on a Cross-Layer Approach for Efficient Video Dissemination in Mobile Multimedia IoT Applications," *Elsevier Computer Communications*, vol. 45, pp. 21–31, 2014.
29. F. Al-Turjman, H. Hassanein, and M. Ibnkahla, "Towards Prolonged Lifetime for Deployed WSNs in Outdoor Environment Monitoring," *Elsevier Ad Hoc Networks Journal*, vol. 24, no. A, pp. 172–185, Jan. 2015.
30. F. Senel and M. Younis, "Relay Node Placement in Structurally Damaged Wireless Sensor Networks via Triangular Steiner Tree Approximation," *Comput. Commun.*, vol. 34, no. 16, pp. 1932–1941, Oct. 2011.
31. R. Rahmatizadeh, S. Khan, A. P. Jayasumana, D. Turgut, and L. Boloni, "Routing Towards a Mobile Sink Using Virtual Coordinates in a Wireless Sensor Network," in *Proc. of the IEEE ICC*, pp. 12–17, June 2014.
32. Mathworks, Simulink, SIMUTOOLS2010.8727. Available at https://www.mathworks.com/products/simulink/. DOI:10.4108/ICST.SIMU TOOLS2010.8727.

8

FOG-BASED CACHING AND LEARNING FOR INFORMATION-CENTRIC NETWORKS

FADI AL-TURJMAN

8.1 Introduction

The increasing demand for highly scalable and efficient distribution of content has motivated the development of future Internet architectures based on named data objects (NDOs), for example, webpages, videos, documents, and other pieces of information. The approach of these architectures is commonly called *information-centric networking* (ICN). In contrast, current networks are host-centric where communication is based on named hosts, for example, web servers, PCs, laptops, mobile handsets, and other devices. Information-centric networks serve as a content-based model that focuses on clients' demands disregarding of the data's address or the origin of distribution. ICN is the next generation model for the Internet that can cope with the user's requests/inquiries regardless of their data-hosts' locations or nature. The current Internet model is suffering from the exchange of huge amounts of data while still relying on the very basic network resources and IP-based protocols. Meanwhile, ICNs promise to overcome major communication issues related to the massive amounts of distributed data in the Internet. ICNs adopt a content-centric architecture which focuses more on the networked data itself rather than the metadata. This kind of network architectures are known usually by the term *content-oriented networks* (CONs) [1]. Luckily, these CON architectures match with the emerging communication trend that aims at exchanging big data over tiny and energy-limited wireless sensor networks (WSNs) in order to realize numerous attractive projects such as the smart-planet and the Internet of things [2–4]. Thus, a new platform is needed to meet these requirements. A new platform, called *fog computing* [5], or simply *fog* because "fog" is a cloud close to the ground, is proposed to address the aforementioned requirements. Fog is a mobile edge computing (MEC) that puts services and resources of the cloud closer to users to be facilitated in the edge networks.

Unlike cloud computing, fog computing enables a new breed of light applications and services that can run at particular edge networks, such as WSNs. In order to enable WSNs to support this trend in communication and function in a large-scale application platform, such as the fog computing, we proposed the cognitive framework in our previous work [6]. In [6], an information-centric scheme is proposed for WSNs using *cognitive* (intelligent) in-network devices that makes dynamic routing decisions based on specific knowledge and reasoning observations in WSNs. *Knowledge* representation using the <*attribute, value*> pair and *Reasoning* using analytic hierarchy process (AHP) techniques are employed at the cognitive device in order to decide on the best data route. AHP is applied on quality of information (QoI) attributes in next-generation WSNs such as reliability, delay, and network throughput observed over the communication links/paths [7,8]. This cognitive information-centric sensor network (ICSN) framework is able to significantly outperform the *noncognitive* ICSN paradigms. However, this cognitive ICSN framework did not consider yet the in-network caching feature. Caching in multitudes of nodes in ICNs has a pivotal role in enhancing the network performance in terms of reliability and response time. In this chapter, we propose the use of value of sensed information (VoI) cache replacement strategy. It identifies the most suitable data to be replaced in order to maintain prolonged data availability periods while enhancing the network performance.

Unluckily, traditional cache replacement strategies in the literature [9,10] has been designed mainly for IP-based computer networks and data centers, which have distinct characteristics in locating data from the envisioned next generation networks such as the light-weight ICSNs. In fact, choosing the most appropriate caching strategy can have significant implications on the overall network performance in terms of the data *publishers' load*, *hit-ratio*, and *time-to-hit* metrics. Several works in the literature have addressed each of these metrics separately. However, since a single ICSN network can serve numerous kinds of applications/users with varying design requirements, we believe in the necessity of a generic dynamic utility function that can consider all the aforementioned metrics while setting different weights for each depending on the ICSN application.

To this end, we provide a novel utility function that sets a value to each cached data item in an ICSN framework. This utility function can determine which data item to drop from the cache while experiencing limited hardware resources for caching. Furthermore, we provide a cache replacement strategy that depends on the VoI in choosing the most appropriate data to be replaced in the cache. We compare our VoI approach against three dominant cache replacement approaches: Node functionality-based caching (FC), content-based caching (CC), and location-based caching (LC) with regard to various performance metrics under a variety of parameters including cache size, data popularity, in-network cache ratio, and network connectivity degree.

The rest of the sections in this chapter are organized as follows. A literature review on caching approaches in ICSNs is provided in Section 8.2. In Section 8.3, we introduce our ICSN-specific system model based on which we build our proposed caching strategy. Section 8.4 explains the proposed VoI approach and utility function. Section 8.5 presents extensive simulation results of the VoI in comparison to other caching strategies. And finally, we conclude our work in Section 8.6.

8.2 Related Work

At the core of the fog paradigm, data has to "live" near its requesters. This is the task of caching schemes. Thus, efficient caching mandates two properties: (1) Ensuring an updated copy of the requested data resides at entities close to the region of interest (in terms of reachability latency) and (2) that copy remains "live" for as long as interest in it exists. Caching is coupled with routing and naming architectures. For example, in DONA, the coupling of naming tuples—Owner public key with data Label—enables in-network caching for any entity in the network that could hold a valid copy. Architectures differ in deciding which entity allows a copy of the data and the basis upon which it would retain it. A recent effort in age-based caching argues for a twofold metric for caching a NDO replica. If it resides at a network edge, or has higher popularity, it will remain cached for a longer duration. Entities which hold replicas will collaborate in "tuning" the age counter to manifest such factors. A core disadvantage at many caching protocols is the inherent need for book-keeping. The resulting message exchange overhead cannot scale to the Internet and yet claim efficiency. We are bound to analyze caching schemes under the following conditions: (1) Communication overhead per NDO; (2) storage requirements; (3) NDOs with different priorities; and (4) granularity in assessing request frequency, types, locations, and so on. We review the different caching techniques in ICSNs and identify the techniques that are best suited for the caching decisions. Accordingly, we categorize the in-network caching in ICSNs into the following categories: (1) Location-based caching, (2) content-based caching, and (3) functionality-based caching.

8.2.1 Location-Based Caching (LC)

Chai et al. in [11] have argued against caching the data everywhere in ICSNs and recommended caching less in order to achieve better network performance. Their caching policy claims that data shall be only cached at the nodes having the highest probability of getting a cache-hit on the data delivery path. Eum et al. [12] have proposed an ICSN architecture, called Cache Aware Target idenTification (CATT). This architecture assumes a topology-aware caching policy, where a node on a downloading path is selected for caching as long as it has the highest connectivity-degree based on the geographical location of this node. However, this can make this kind of nodes sort of a geographical bottleneck in the network. Meanwhile, authors in [13] have investigated the performance of topology-based replica placement on Internet router-level topology and found out that the router-level fan-out placement is almost as good as the greedy placement of replica. Moreover, they found that a fan-out based replica placement method needs to be carefully designed to be efficient in content oriented architectures. Works in [11–13] base on the node degree or node fan-out for replica placement, but these methods cannot be universal because node degree–based solutions cannot be good solutions if most of the nodes have similar, relatively low degree, or fan-out.

Bhattacharjee et al. [14] have considered the use of various self-organizing or active cache management strategies in which nodes make globally consistent decisions about caching and revealed that in many cases, these self-organizing caching schemes yield better average delays than traditional approaches (cache at transit nodes), using much smaller per-node caches. A selective neighbor caching approach that selects an appropriate subset of neighboring proxies that minimizes the mobility costs in terms of expected average delay and caching costs has been proposed by Li et al [15]. This approach is based on proactively caching data requests and the corresponding metadata to a subset of proxies that are one hop away from the proxy. Authors in [16] suggest a probabilistic approach for ICNs. They claim that the probability of a file being cached should be increased as it travels from source to destination by considering the following parameters: (1) The distance between source and current node, (2) distance between destination and current node, (3) time-to-live for the routed data content, and (4) the time-since-birth. The authors also suggest redundancy in caching on a single path between source and distention. However, this degrades the ICN performance dramatically while experiencing limited caching spaces. Moreover, in [16], authors assume that all the network nodes have the capability of caching, which is not the case in practice with fog systems. The proposed approach is weak as well due to considering static data requests' frequency from a subnet where that data can exist. Nevertheless, we believe caching should be based on dynamic frequencies and location-independent.

8.2.2 Content-Based Caching (CC)

Content-based caching is another candidate category for caching ICNs, in which the data replacement decision is taken based on the content of the exchanged data. For example, authors in [17] propose an autonomic cache management architecture that dynamically (re)assigns data items to in-network caches. Distributed managers make (re)placement decisions, based on the observed data request patterns such as their popularity, in order to minimize the overall network traffic. Also in [17], the authors suggest that every cache manager should decide in a coordinated manner with other cache managers whether or not to cache an item. This approach assumes that every cache manager has a holistic network wide view of all the cache configurations and relevant request patterns. And thus, it adapts depending on the volatility of the user requests. It is evident that the network-wide knowledge and cooperation give significant performance benefits and reduce significantly the time to convergence, but at the cost of additional message exchanges and computational overhead.

Meanwhile, the main objective of the work in [18] is to minimize the Internet service provider (ISP) traffic and accessed in-network hops/devices by caching frequently requested data at ISP-specific routers. The main problem addressed here is to provide effective caching strategies for these routers to coordinate their data replacement based on their content. Guided by optimal replica placement, the authors have presented two popularity-based caching algorithms. However, this work may not be practical as the authors have assumed only one gateway in an ISP network. Cho et al. [19] have proposed a content caching approach, called WAVE, in which the cache size is adjusted based on data popularity. In WAVE, an upstream node recommends the number of chunks to be cached at its downstream node, which increases exponentially as the data requests increase in order to reduce communication and cache management overhead. WAVE distributes content chunks toward the network edge (from where the data requests come) considering the content popularity and distance relation. However, the different sizes of data chunks have not been considered in this reference. An age-based distributed cache approach aiming at reducing the data publisher load and in-network delay for ICSNs has been proposed in [20]. This approach provides a lightweight cooperative mechanism to control where data contents' ages are dynamically updated implicitly. It spreads popular contents towards the edge of ICSN and meanwhile eliminates the unnecessary replicas at the intermediate ICSN nodes. Yet this approach suffers from maintaining highly dynamic contents, and thus, nodes which are far away from the server may experience long time periods to refresh their contents.

8.2.3 Node Functionality-Based Caching (FC)

To unleash the full potential of ICSNs, the role (function) of the in-network caching node shall be taken into consideration; considering which content to cache at the management/control level rather than guessing it at the data level. Authors in [21] remarked the side-effects of delegating the caching decision to the data level and propose a specific approach to handle data caching at the control level. The proposed approach can be tuned to make sort of a balance between the benefits and cost overhead. However, it is only applicable at a small scale and it may not accommodate the massive amounts of data contents in the Internet. The authors in [22] proposed *LocalGreedy* algorithm for caching in ICNs. They consider a cache cluster consisting of number of leaf nodes which are either directly connected or indirectly via first node as a common parent somewhere along the path to the root node. Furthermore, they assume interlevel cache cooperation, which is basically a special case where content can only be fetched from a specific node, called the parent node, and not from any peer node in the network. Authors claim that a strong conceptual similarity emerges when the notion of "access cost" is adopted. The access cost is represented either by the incurred latency while fetching content from remote caches, or by the bandwidth consumed when retrieving content from a peer node or video head end, depending on the scenario of interest. However, this approach necessitates a global knowledge of the in-network nodes' capability and this contradict with the fog vision. In [23], the authors investigated the trade-off between caching data contents in a distributed IP-based networks and the new emerging ICSN architectures in fog systems such as the content-centric network (CCN). They applied their study on the real traffic mix resulting from several functional resources nowadays such as the web, file sharing, and multimedia streaming. It has been demonstrated that caching videos in routers offers more cache hits. Nevertheless, the other types of content would likely be more efficiently handled in very large capacity storage devices in the core of the network. And thus, this kind of caching is not efficient in ICSNs.

In recent times however, the Internet has progressed towards an information-centric sensor networking paradigm, where the focus is on delivering named blocks of data to users at the network edge rather than establishing end-to-end connections to the web server. So the design of the cache replacement policy in ICSNs must be a dynamic one, based on the user's request trends and the application on hand. In this chapter, although we need to use a content-centric approach, the same cache replacement approaches cannot be applied to an ICSN. This is because of the unique resource constraints of the sensor network, the uncertainty of the wireless medium, and the need to be aware of user requirements in the ICSN architecture. The resource limitations of the sensor network nodes include limited power supply, storage space, and heterogeneity in terms of the sensors used and the node functions. In addition, the same content (sensed data) cannot be replicated into multiple caches without associating them with location, because the sensed information may be different in different parts of the network, and it may change over time too, which is unlike the case of ICSNs. This makes cache replacement trickier in information-centric sensor networks. In addition, the replacement policy should take into account the type of user requests coming to the network, the sensor node availability at different locations (as nodes eventually die out), and also the sensing duration for different sensors on board the sensor nodes. In this chapter and unlike other related work, dynamic caching decisions is made based on specific knowledge and reasoning observations in the network. Based on these cognitive elements/observations, we prioritize between the cached contents in a hierarchical storage system (i.e., in a multilevel caching). Accordingly, we provide a novel utility function that sets a value to each data item based on application-specific metrics, such as required quality of communication channel, delay, and data age. This makes our proposed VoI approach able to cope with fog networks as a new trend in communication.

8.3 System Models

In this part of our work, we elaborate on our ICSN network model in addition to its corresponding data popularity, age, and delay models.

8.3.1 ICSN Network Model

The main components in our ICSN network model are listed as follows. Sensor nodes (SNs) to sense the environment and capture physical changes in the surrounding environment, and report these changes via relay nodes (RNs) or local cognitive nodes (LCNs). They interact with the RNs and LCNs as shown in Figure 8.1. LCNs have elements of cognition, that is, knowledge, reasoning, and learning, which help in interpreting common requests and queries' responses. They interact with SNs, RNs, and the sink. RNs forward the data

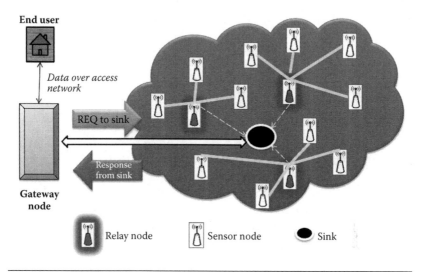

Figure 8.1 Network model with sensor nodes and cognitive nodes.

received from SNs to the sink or any neighboring local cognitive nodes in response to received requests from the user. The sink is where all the collected data is delivered. The sink node is also enhanced with cognitive elements to be more intelligent in managing the network performance based on data traffic type and it is called a global cognitive node (GCN).

The type of data traffic that an ICSN paradigm can handle is categorized into one of following: (1) *Type I: On-Demand*, (2) *Type II: Periodic*, and (3) *Type III: Emergency*. Each of these types is associated with a different quality of information (QoI) value on the cached data, based on the ICSN application. We select the network *Reliability (R)*, *Latency (L)*, *Energy (E)*, and *Throughput (T)* as the four main attributes, whose combined value decides the QoI of the cached data. We are not using the absolute values for these attributes. Instead, we associate priorities with each of these attributes for every request type, and make these priorities decide the importance of the absolute value of the attributes, as shown in Table 8.1. The values in Table 8.1 indicate the associated priority with each attribute. Number 1 indicates the highest priority and number 3 indicates the lowest. The "x" in Table 8.1 indicates a "don't care" condition. This means that there are no strict requirements on the value of the QoI marked with an "x," and its value does not affect the caching decision.

Table 8.1   QoI Attribute Priority for Different Data Traffic Types

	QUALITY OF INFORMATION (QoI)—ATTRIBUTES			
REQUEST TYPE	LATENCY (L)	ENERGY (E)	RELIABILITY (R)	THROUGHPUT (T)
Type I: On-Demand	x	3	1	2
Type II: Periodic	1	2	4	3
Type III: Emergency	1	1	x	2

8.3.2 Delay Model

Different sensors have different durations for which they need to be exposed to the environment, so that they can capture the sensed readings accurately. This affects the duration of the on-time of the sensor node, which in turn affects the lifetime of the sensor node [24]. In order to prolong the lifetime of the sensor node, it is useful to store the sensed data for longer when the delay involved in acquiring the reading is more. This is called the *sensing delay*. In addition, if data has to be propagated from sensor nodes to LCNs every time data is requested, it would add to the propagation delay of the data, especially if the sensor nodes are located far away from the sink. Thus the delay components we consider are the sensing delay δ and the propagation delay τ. We also limit the number of hops (n) within which the data has to be delivered to the sink to 6, so as to avoid unnecessary wastage of energy by involving multiple nodes in the data transmission. Accordingly,

$$\tau \propto n, \quad \text{where } n < 6 \tag{8.1}$$

$$\delta \propto \max(d_1, d_2, d_3, \ldots d_k) \tag{8.2}$$

where k is the total number of sensors available on board the sensor node, and d_i represents the fixed sensing delay value of the sensor type i (see Table 8.1). Thus, the sensing delay is a function of the maximum delay from among the sensor types that have been activated to provide fresh data. Putting these two delays together, the total delay (Δ) involved in delivering freshly sensed data to the sink is a combination of the sensing and propagation delay, given by Equation 8.3:

$$\Delta = \tau + \delta \tag{8.3}$$

8.3.3 Age Model

Our age model makes use of the following two conditions to decide what content should be dropped from the cache.

The first is the based on the periodicity of the periodic request (Type I traffic), and the second, when the node's cache is full. We make use of the periodicity of the periodic request, because freshly sensed data has to be provided at the start of each periodic request cycle. Thus, when the cache is full at the end of one periodic request cycle, old data can be discarded from the cache. Thus, the age of a sensed attribute–value pair is represented by its time-to-live (TTL) which is based on the periodicity of the request of each application type. This value is provided to the LCN by the GCN/sink. Since we are not considering the use of historic data, our model implies that cached contents may be refreshed after every periodic time interval, as long as the data is being transmitted to the sink at the end of each cycle [25,26].

$$TTL_{Si} \propto T_{periodic} \tag{8.4}$$

Equation 8.4 represents that the TTL of the sensed information (Si) represented as attribute–value pair, is directly dependent on the periodicity of a request in Type I traffic flow. In case the application requires that the periodic data is stored for a prolonged duration of time, for example, 24 hours, before making a single transmission to the sink, then the cache retention period becomes a function of the transmission cycle's periodicity.

8.3.4 Popularity of On-Demand Requests

Traffic flow generated in response to on-demand requests have been classified as Type II traffic. More numbers of users may be interested in a particular type of sensed data, or a specific sensed data may be requested more numbers of times by one or more users. Such sensor data is said to be popular, and can be retained for longer in the LCN's cache. Thus the popularity of the sensed attribute–value pair is given by Equation 8.5.

$$Popularity_{Si} \propto Re\,q_{Si}/Re\,q_{total} \tag{8.5}$$

where Req_{Si} is the total number of requests for an attribute–value pair received at an LCN, and Req_{total} is the total number of requests received by that LCN within a particular operational cycle. In addition, when sensor nodes start to die out in the network, LCNs should store the data for longer to maintain their availability. When the primary LCN storing such data itself starts to die out, storing the data in neighboring LCNs provides extra storage guarantees and ensure availability of data in the network for longer. This storage requirement based on non-availability of alive sensor nodes is managed by the planning algorithm for data delivery based on the traffic flow in the network and remaining energy at LCNs.

8.3.5 Channel Communication Model

Here we elaborate on the assumed channel model in our wireless communications [27]. The transmission power utilized by the *ICSN* nodes is represented as T_{P_0} and the transmission range between *BS* and an *SN* is represented as T_r. The expression of channel model can be provided as

$$C_M = A\rho T_{P_0} T_r^{-\alpha} \tag{8.6}$$

Where C_M is the transmission power of the *BS*, A is the constant gain factor for power provided by antenna and amplifier gain, ρ is the small scale constant for fading factor, and α is the path loss exponent. The transmission range between the *ICSN* nodes of i and j is denoted as R_{ij} ($i, j = 1, 2, 3, \ldots\ldots, N$). The range for *ICSN* nodes (i and j) and the *BS* is denoted as r_i, and the power transmission for node i is defined as p_i. The link interference is expressed as

$$I_{BS} = \frac{A\rho T_{P_0} T_r^{-\alpha}}{A\rho T_{P_i} r_i^{-\alpha}} \tag{8.7}$$

8.4 VoI Cache Replacement

For cache replacement in ICSNs, we need to ensure that we choose data appropriately for storage based on the following criteria: First, data that takes longer to sense should be stored for longer to conserve the sensor node's energy. Second, data storage must be a function of the periodicity of the requests based on the traffic type. This will help to store data till fresher data is available, and in servicing requests for different traffic types in a timely manner. Last, value of the data based on its age, that is, if temperature in a region has changed considerably from the last time it was sensed, then the cached information is stale and does not provide correct information. Hence the freshness of data is also an important criterion when servicing requests for data on demand. Since these criteria are known and fixed, the cache replacement plan can be programmed into the LCN.

We propose a VoI based cache replacement strategy for the LCNs in an ICSN. Our cache replacement approach adopts the aforementioned system models to achieve an efficient cache management strategy that can handle the following three types of content:

1. Delay-based content: The delay sensitivity of the cached data content is a measure specified by the requesting user to indicate how long the consumer is willing to wait for it. Examples of delay-sensitive data can be found in applications serving areas of emergencies (e.g., disaster or health emergency).
2. Demand-based content: This is a measure of the data popularity which is specified by the frequency of requesting a specific data.
3. Age-based content: Some contents are more sensitive to aging. For instance, if a user requests information about the traffic updates for the coming 30 minutes, then any related content that does not cover this time interval is useless.

Accordingly, the VoI cache management approach employs three parameters to set a *value* VoI_{Si} per sensor node S_i reading. This value is dependent on the history of the data content within each operational round. At the beginning of each round and based on the aforementioned models, all content resets its VoI_{Si} value according to the following function:

$$VoI_{Si} = \alpha * \Delta + \beta * TTL_{Si} + \gamma * Popularity_{Si} + \lambda * 1/I_{BS} \quad (8.8)$$

where α, β, γ, and λ are the tuning parameters that are specified based on the traffic type and the user requests. The strength of VoI is mainly in its ability to prioritize based on the targeted ICSN application. Therefore, to improve the basic priority caching method, the weights of each of the parameters can be adjusted to find the most efficient approach. The delay sensitivity parameter serves in assuring the least delay. The popularity parameter is important, as it takes into consideration the most frequently demanded data packets. The packet age value is also vital since it considers packets in the cache that have not been used for long time and replaces them with more relevant data. In the following, Algorithm 8.1 provides the steps to be executed by each node if its cache is full to drop data with the least VoI_{S_i}.

Algorithm 8.1 Drop Least VoI_{S_i}

1. **Function VoI** (*content*)
2. **Input**
3. *content: A content item within the ICSN*
4. **Begin**
5. **for** each LCN node, **do**
6.   **for** each duty cycle, **do**
7.   **Set** *value* of each VoI_{S_i} in the cache based on Equation (8.8)
8.   **if** *cache_full*
9.   Check history of user requests
10.   Drop the data content of the least VoI_{S_i}
11.   **End if**
12.   **End for**
13. **End for**
14. **End**

In the above algorithm, elements of cognition are implemented at the LCNs. These elements are *reasoning* and *learning* elements.

8.4.1 Learning

Learning is used in our VoI approach in order to determine the most appropriate paths towards the GCN that satisfy the fog network requirements. This cognition element uses a direction-based heuristics to determine the data delivery path through RNs that lie in the direction of the GCN. Hence, each time a LCN has to choose the next hop, the direction-based heuristic eliminates RNs that increase the distance between the current RN and GCN. This information is stored in the LCN for use in the next transmission rounds. Thus the direction-based heuristic, along with feedback from the network about the chosen paths helps the LCNs to learn data delivery paths to the sink, as the network topology changes.

Example 8.1: Assume S_1 and S_2 have data to be sent to destination nodes D_1 and D_2. R_n are all the available relays toward the destination. Out of these relays, it is determined that R_5 as shown in Figure 8.2 has the lowest link outage probability to D_1 and D_2. Therefore, S_1 initiates routing data to R_5. Meanwhile, S_2 also forward a high traffic of data to R_5 (depicted by solid paths in Figure 8.2). When multiple source nodes start routing their data to R_5 as well, the route to R_5 may get congested. A cognitive network with *learning* capabilities will be able to identify the congestion at R_5 (by observing the decrease in throughput). Sharing this observation with neighboring nodes, the cognitive fog network (or ICSN) would be able to respond to the congestion proactively, by routing the data through a different path involving nodes R_4, R_8, and R_9 as shown in Figure 8.2(b).

8.4.2 Reasoning

In the VoI approach, we assume a modified version of the analytic hierarchy process (AHP) [22] for implementing the reasoning element of cognition in the fog network. AHP supports multiple-criteria decision making while choosing the data path. For example, if we have a delay-sensitive data, the node which provides the lowest latency will be chosen even though it might degrade other metrics, such as the network energy or throughput. If two next hops guarantee the same latency then the next attribute to compare will be energy, and then throughput, assuming that energy is the next desired attribute in the fog network. AHP provides a method for pairwise comparison of each of the attributes and helps to choose the node that can provide the best network performance on the long run. The following example has more details on the utilized AHP.

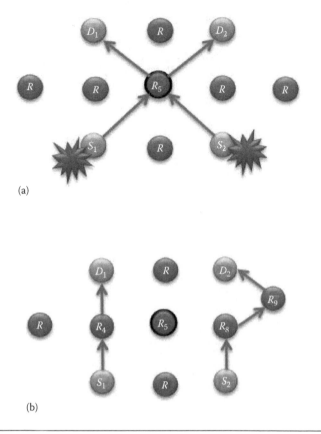

(a)

(b)

Figure 8.2   (a) Classical routing in a sensor network. (b) Cognitive routing in Fog.

8.5 Performance Evaluation

In this section, we provide initial performance evaluation results for the VoI-based cache replacement technique, which we have compared with node functionality-based caching (FC), content-based caching (CC), and location-based caching (LC) techniques using NS3, a discrete event simulator. The caching schemes: FC, CC, LC, and VoI, are executed on 600 randomly generated wireless heterogeneous network topologies in order to get statistically stable results. The average results hold confidence intervals of no more than 5 percent of the average values at a 95-percent confidence level. We make use of the cache hit ratio to compare the performance of the different cache replacement strategies. Cache hit ratio is defined as the ratio of the number of times requested data was found in the cache divided by the total number of times data was requested from the cache. The storage cache is implemented as a single storage level in one case (L1 cache) and as a hierarchy of two storage levels in another case (L1 and L2 cache). Simulation results are compared for VoI, LC, CC, and FC replacement techniques. These simulations were run at cache sizes ranging from 10 to 100, and the simulations end after serving 1000 packet requests. There are 100 different requests from which the packet requests are randomly generated.

8.5.1 Performance Metrics

To compare the performance of the proposed VoI approach, we track ICSN-specific metrics to achieve qualitative conclusions for the targeted in-network caching problem. We simulate the performance of an ICSN network with the detailed physical layer NS3 built-in parameters so that we achieve realistic simulation instances. The four considered performance metrics are as follows:

- Cache-hit ratio: Is simply the fraction of time a request arrives at a node to which that cache is attached but does not contain the requested data item. It is the average hitting ratio over all the in-network caches. We preferred to look at average time to hit data and hitting ratio more than publisher load, but we generally expect publisher load to improve as the other metrics improve as well.
- Time-to-hit data (TTH): Is found by simply logging all the total costs of the request and response paths incurred by every sensor node. Ideally, ICSN is supposed to minimize the total average time to hit data per request.
- In-network latency (delay): This metric represents the end-to-end delay as described above. Note that we differentiate between latency to hit data and in-network latency since the two metrics may differ because of mobility or disruption conditions.
- Average request per publisher (ARP): This metric is measured in number of data request per hour (requests/hour) and it represents the average load per publisher in an ICSN paradigm. We track publisher load by monitoring the total fraction of data requests that had to be satisfied by a data publisher.

8.5.2 Simulation Parameters

Many of the ICSN paradigm parameters have to remain fixed while our simulation instances are generated. In particular, the parameters of our simulation are as follows:

- Percentage of nodes with caches (PoC): This parameter is our primary method for controlling the extent of caching in our ICSN. By varying this parameter, we can study the sensitivity of metrics like time-to-hit data to the caching extent.
- Connectivity level (degree): It represents how tightly connected the ICSN network is. We use the connectivity matrix, based on our described communication model in Section 8.3.
- Data Popularity: It indicates how frequent a specific data content is requested. This metric is measured in percentage with respect to other requested data contents. This parameter is represented by a single Poisson process parameter in order to give the content replacements per time unit.

8.5.3 Simulation and Results

The following figures depict the achieved results. Our first objective is to confirm that increasing the extent of caching in ICSNs, in terms of both size and number of levels, will reduce time to meet data for all cache policies.

According to Figure 8.3, we can deduce that the VoI is not efficient in level one cache, however, FC and LC cache replacement techniques perform equally well in this scenario. In Figure 8.4, where we have two levels of caching (one on LCN and the other on RN) we find out that VoI surpasses the other two replacement techniques. Since we have only one type of the considered requests, there is minimum performance gain when the cache size is increased beyond 30 megabytes.

The next set of simulations' figures (Figures 8.5 and 8.6) are set to analyze the performance of the cache replacement strategies as the number of requests that a given network needs to serve increases from 500 to 5000 requests/hour. The cache size is set to be 100 megabytes and the number of request types is fixed at four.

From the figures we can deduce that the advantage held by the VoI replacement technique is that it replaces data based on user requirement

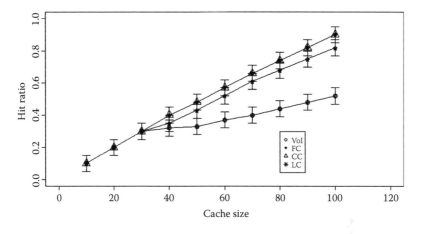

Figure 8.3   Cache size vs. the hit ratio with one-level caching.

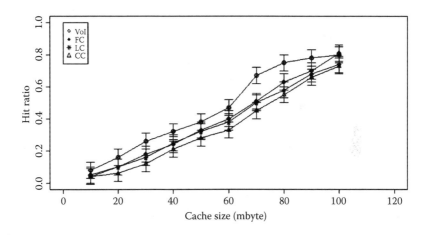

Figure 8.4   Cache size vs. the hit ratio with two-level caching.

and VoI of the data. Other replacement techniques are only concerned with the match of the data requested packet number, without considering the age of the data, its popularity, or the delay associated with sensing and transmitting it to the sink. Nevertheless, the use of VoI replacement technique puts into consideration the age of data, VoI, and the popularity of the information. New information replaces old ones, unlike other technique that only find a number match irrespective of their age. Based on this we proposed the use of a two-level cache, one on LCN and the other on RN, in order to reduce the complexity of computation. We can employ the use of the VoI base replacement technique on LCN and FC or LC on RN. We can set the size of level 1 and 2 to 100 and the packet

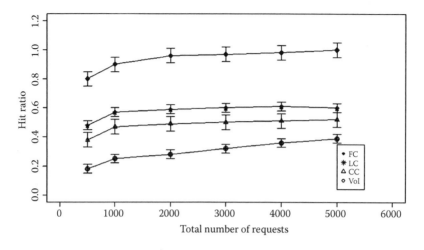

Figure 8.5 Total number of requests vs. the hit ratio with one-level caching.

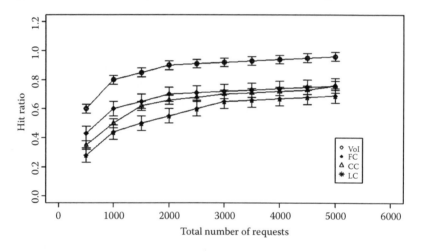

Figure 8.6 Total number of requests vs. the hit ratio with two-level caching.

request can be 10000. We can set the size of level 1 and 2 to 100 mega-bytes and the number of packet requests to 10000.

From Table 8.2, we can see that for the two levels of cache, the best possible combinations are VoI-based replacement strategy at L1 cache and VoI or LC based replacement strategy at L2 cache. Despite having a good hit ratio sometimes, FC is not reliable to serve the user needs considering the age of data and the delay required to retrieve the required data because the decision making is only at LCNs. Implementing the VoI replacement technique at LCNs can greatly help save resources, especially when a cache hit is found on the first level of the cache.

Table 8.2 Two-Level Caching Comparison

L1 CACHING POLICY	L1 HIT RATIO	L2 CACHING POLICY	L2 HIT RATIO	TOTAL HIT RATIO
VoI	0.811542	LC	0.81743	0.81542
VoI	0.547	FC	0.7792	0.6194
VoI	0.899	VoI	0.0099	0.81743
LC	0.754	VoI	0.398	0.6837
LC	0.9	FC	0	0.71818
LC	0.802	LC	0.49	0.75125
FC	0.9	LC	0	0.71818
FC	0.9	VoI	0	0.61818
FC	0.9	FC	0	0.65125

Figures 8.7 and 8.8 below represents the findings from the simulation experiment of the two-level cache. From both figures, the extend of cache availability increases proportionally. According to Figure 8.7, we observe that the overall time to meet data, which is our main performance metric, is reduced in all performance policies. However, the VoI policy performs best at higher proportions of nodes attached to caches. On the contrary, Figure 8.8 shows that there is an increase in the data hit for all the approaches, and hence, we conclude that the VoI is better due to its ability to replace the most relevant data according to an ICSN-specific set of attributes.

In Figures 8.9 through 8.11, connectivity level (degree) is the examined parameter. From Figure 8.9, we deduce that there is an

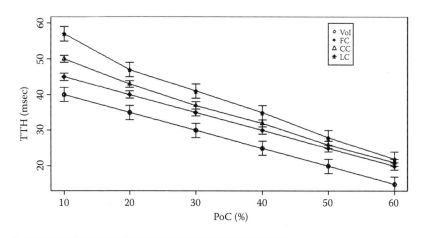

Figure 8.7 Time to hit ratio vs. percentage of nodes with caches (with connectivity degree = 30).

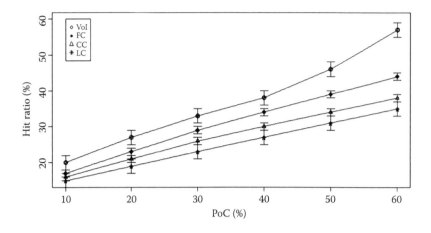

Figure 8.8 Hit ratio vs. percentage of nodes with caches (with connectivity degree = 30).

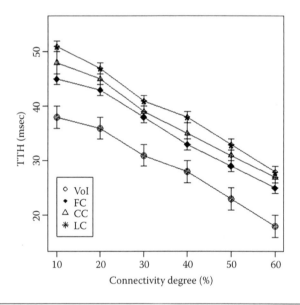

Figure 8.9 Time to hit ratio vs. the connectivity degree percentage.

increase in time to hit data as the ICSN connectivity increases in all the approaches. However, we notice that VoI is less dependent on the network and hence better than the other two approaches. VoI is rather more dependent on the data type, which is a highly desired property in ICSN network. Figure 8.10 shows the data hit performance against a varying network connectivity degree, and while applying the VoI scheme, we notice that the data hit increases exponentially while

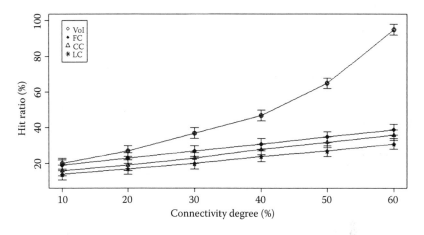

Figure 8.10   Hit ratio vs. the connectivity degree percentage.

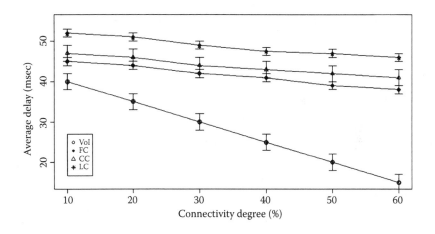

Figure 8.11   Average in network delay vs. the connectivity degree percentage.

the network connectivity increases. Nevertheless, the data hit of the other two approaches increases linearly. Moreover, Figure 8.11 shows that VoI is the best in terms of delay. This can be attributed to the application of the delay factor while deciding which data to replace. Figure 8.12 shows the effect of data popularity in terms of publisher load. The VoI tops LC and FC as the popularity metric increases. This is a very desirable property in the ICSNs.

The VoI-based technique would be suitable for the ICSN approach, as we propose to use a named data association for the sensed data, such as attribute–value pairs, and the cache size can be decided based

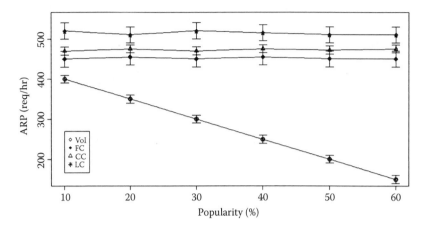

Figure 8.12   Publisher load vs. the data popularity.

on the different types of user requests that the network is expecting to serve. We can expect that the users are more satisfied with the response received from the LCN, as it retains information in its cache based on both data popularity and various parameters that affect gathering sensed information and the energy involved in doing so, as the network scales up to larger sizes.

8.6 Conclusions

The VoI-based technique is suitable for use in ICSNs that use named data association for the sensed data, and support node mobility. We can expect that the users are more satisfied with the response received from the local cognitive nodes (LCNs), as they retain information in their cache based on both data popularity and various parameters that affect gathering sensed information and the energy involved in doing so, as the network scales up to larger sizes. Further, the VoI cache replacement strategy helps in graceful degradation of the network, as cached data can be provided from LCNs even after the sensor node death. Moreover, this work further investigates and evaluates the impact of varying network loads and inter-LCN communication on the effectiveness of the cache replacement strategy.

Consequently, VoI realizes the delay-tolerant data requirements in edge networks. It provides a dynamic data replacement policy that is based on the consumers' requesting trends. It maximizes the consumer's gain according to delay limit, data popularity, and time-to-live metrics. Moreover, it maximizes the gain of the data publishers by reducing their load. Our presented simulation results in this work show the efficiency of VoI when compared to other approaches such as FC and LC. It exhibits superior performance in terms of publisher load, end-to-end delays, time-to-hit data, and hit ratio.

References

1. B. Ahlgren, C. Dannewitz, C. Imbrenda, D. Kutscher, and B. Ohlman, "A Survey of Information-Centric Networking," *IEEE Communications Magazine*, vol. 50, no. 7, pp. 26–36, 2012.
2. SmartSantander, Future Internet Research and Experimentation. [Online.] Available at http://www.smartsantander.eu.
3. IBM | A Smarter Planet | Smarter Cities. [Online]. Available at http://www.ibm.com/smarterplanet/us/en/smarter_cities.
4. F. Al-Turjman, "Cognition in Information-Centric Sensor Networks for IoT Applications: An Overview," *Ann. Telecommun.*, pp. 1–16, 2016. DOI: 10.1007/s12243-016-0533-8.
5. F. Bonomi, "Connected Vehicles, the Internet of Things, and Fog Computing," VANET 2011, 2011.
6. G. T. Singh and F. M. Al-Turjman, "A Data Delivery Framework for Cognitive Information-Centric Sensor Networks in Smart Outdoor Monitoring," *Elsevier Computer Communications*, vol. 74, no. 1, pp. 38–51, 2016.
7. F. Al-Turjman, A. Alfagih, W. Alsalih, and H. Hassanein, "A Delay-Tolerant Framework for Integrated RSNs in IoT," *Elsevier Computer Communications Journal*, vol. 36, no. 9, pp. 998–1010, May, 2013.
8. F. Al-Turjman, H. Hassanein, and M. Ibnkahla, "Efficient Deployment of Wireless Sensor Networks Targeting Environment Monitoring Applications," *Elsevier Computer Communications Journal*, vol. 36, no. 2, pp. 135–148, Jan. 2013.
9. A. Chankhunthod, P. Danzig, C. Neerdaels, M. Schwartz, and K. Worrell, "A Hierarchical Internet Object Cache," in *Proc. of USENIX*, 1996.
10. M. Gritter and D. R. Cheriton, "TRIAD: A New Next-Generation Internet Architecture," Stanford University, Stanford, CT, July 2000.
11. W. K. Chai, D. He, I. Psaras, and G. Pavlou, "Cache 'Less for More' in Information-Centric Networks (Extended Version)," *Elsevier Computer Communications*, vol. 36, no. 7, pp. 758–770, 2013.
12. S. Eum, K. Nakauchi, Y. Shoji, N. Nishinaga, and M. Murata, "CATT: Cache Aware Target Identification for ICN," *IEEE Communications Magazine*, vol. 50, no. 12, 2012.

13. P. Radoslavov, R. Govindan, and D. Estrin, "Topology-Informed Internet Replica Placement," *Proceedings of WCW'01: Web Caching and Content Distribution Workshop*, 2001.

14. S. Bhattacharjee, K. L. Calvert, and E. W. Zegura, "Self-Organizing Wide-Area Network Caches," in *IEEE Infocom*, pp. 752–757, 1998.

15. X. Vasilakos, V. Siris, G. Polyzos, and M. Pomonis, "Proactive Selective Neighbor Caching for Enhancing Mobility Support in Information-Centric Networks," in *Proc. of the ICN Workshop on Information-Centric Networking*, New York, pp. 61–66, 2012.

16. I. Psaras, W. K. Chai, and G. Pavlou, "Probabilistic In Network Caching for Information-Centric Networks" in *Proc. of the 2nd Edition of the ICN Workshop on Information-Centric Networking*, pp. 55–60, 2012.

17. V. Sourlas, P. Flegkas, L. Gkatzikis, and L. Tassiulas, "Autonomic Cache Management in Information Centric Networks," in *Proc. of the IEEE Network Operations and Management Symposium (NOMS)*, 2012.

18. J. Li, H. Wu, B. Liu, X. Wang, Y. Zhang, and L. Dong, "Popularity Driven Coordinated Caching in Named Data Networking," pp. 200–211, 2012.

19. K. Cho, M. Lee, K. Park et al., "WAVE: Popularity-Based and Collaborative In-Network Caching for Content-Oriented Networks," in *Proceedings of INFOCOM WKSHPS*, pp. 316–321, 2012.

20. Z. Ming, M. Xu, and D. Wang, "Age-Based Cooperative Caching in Information-Centric Networks," *Int. Conf. on Computer Communication and Networks (ICCCN)*, 2014.

21. W. Yaogong, K. Lee, B. Venkataraman et al., "Advertising Cached Contents in the Control Plane: Necessity and Feasibility," in *Proc. INFOCOM Workshop on Computer Communications*, 2014.

22. S. Borst, V. Gupta, and A. Walid, "Distributed Caching Algorithms for Content Distribution Networks," in *Proc. of the IEEE INFOCOM*, 2010.

23. C. Fricker, P. Robert, J. Roberts, and N. Sbihi, "Impact of Traffic Mix on Caching Performance in a Content-Centric Network," in *INFOCOM Workshops*, pp. 310–315, 2012.

24. F. Al-Turjman, H. Hassanein, and M. Ibnkahla, "Towards Prolonged Lifetime for Deployed WSNs in Outdoor Environment Monitoring," *Elsevier Ad Hoc Networks Journal*, vol. 24, no. A, pp. 172–185, Jan. 2015.

25. A. Al-Fagih, F. Al-Turjman, W. Alsalih, and H. Hassanein, "A Priced Public Sensing Framework for Heterogeneous IoT Architectures," *IEEE Transactions on Emerging Topics in Computing*, vol. 1, no. 1, pp. 135–147, Oct. 2013.

26. F. Al-Turjman, H. Hassanein, S. Oteafy, and W. Alsalih, "Towards Augmenting Federated Wireless Sensor Networks in Forestry Applications," Springer: *Personal and Ubiquitous Computing Journal*, vol. 17, no. 5, pp. 1025–1034, June 2013.

27. F. Al-Turjman, H. Hassanein, and M. Ibnkahla, "Quantifying Connectivity in Wireless Sensor Networks with Grid-Based Deployments," Elsevier: *Journal of Network & Computer Applications*, vol. 36, no. 1, pp. 368–377, Jan. 2013.

9

CONCLUSIONS AND
FUTURE DIRECTIONS

Smart and cognitive environments have emerged as one of the most promising applications of WSNs in the Internet of things (IoT) era because of their information-centric nature and significant impact in everyday activity. The IoT domain has been spread across variety of areas including the future Internet, object identification using RFIDs, and various intelligent techniques to improve the network performance and quality of experience. Toward more efficient IoT implementations, in this work, we proposed and evaluated the use of cognitive sensor networks in large-scale IoT applications. We focused on key design aspects in the cognitive node and network architecture, deployment, and data delivery trends.

9.1 Summary of the Book

We started the work in this book with a comprehensive overview about the use of intelligent decisions in sensor networks, while discussing a brief use case as a proof of concept in Chapter 1. From the results of the case study, we concluded the intelligence capability in IoT. We also suggested that it would be useful to develop a generic knowledge-based cognitive framework that can be applied to sensory applications, wherein the network decisions are based on learning and reasoning. From there on, we focused on investigating the cognitive node architecture, deployment, and delivery techniques in low-cost sensor networks; while considering modern applications in smart environments, such as the Smart Cities project.

In Chapter 2, we proposed a cognitive information-centric sensor network (ICSN) architecture toward introducing the cognition concept in IoT using cognitive nodes deployed at specific locations in the network. These cognitive nodes implemented the features of the network status feedback loop via knowledge representation, reasoning, and learning. A cost-comparison discussion between a typical relay node and the proposed cognitive nodes (CNs) concept showed that the latter were more expensive. And hence, a grid-based deployment plan was proposed for the ICSN in order to maintain the least number of CNs in comparison to the number of relays in the network. Distances between the relays and the CNs were planned based on the probability of a successful data reception at the network nodes. A MATLAB® simulation model based on the IEEE 802.15.4 PHY and MAC layer protocols was also used to verify the impact of varying load and number of the active CNs on the network's performance in terms of latency, reliability, and the instantaneous throughput.

In Chapter 3, we detailed a possible cognitive data delivery framework, and elaborated on the reasoning and knowledge representation techniques utilized at the cognitive nodes. Knowledge was represented using attribute–value pairs, and the analytic hierarchy process (AHP) was used as the reasoning element of the cognition in the ICSN for more quality of information (QoI)–aware decisions. We showed how the CNs put these elements of cognition in use by identifying data delivery paths to the sink from any source in the network. Latency, reliability, and throughput were the attributes used to identify the QoI-associated with data that was delivered to the end user. Based on extensive simulation results, we concluded that by restricting the number of nodes scheduled for simultaneous transmission, and the per-node offered load, the targeted QoI attributes can be maintained at acceptable levels along the data delivery paths. In addition, the network performed very well while responding to varying traffic types and network topologies, thus establishing the suitability of the cognitive ICSN for smart environments in IoT.

In Chapter 4, we presented an Enhanced 2-Phase Data Delivery (E2-PDD) framework for ICSNs, focusing on efficient content access and distribution as opposed to mere communication between data consumers and publishers. We employed an approach of growing eminence, where requests are initiated by consumers seeking particular services that are data-dependent. High-level controllers (HCs) receive the consumers' requests and issue queries to a multitude of data publishers. The publishers in the examined topology included a wide variety of ubiquitous nodes that could be either stationary or mobile, operating under different protocols. In order to consider fundamental challenges in ICSNs designed for dynamic IoT applications such as the mobility and data disruptions, our E2-PDD framework employs low-level controllers (LCs) that act as moderators between the HCs and the data publishers, executing data queries for the top tier and replying back with a set of candidate rendezvous points obtained from a bottom tier. Extensive simulation results had been used to evaluate the proposed E2-PDD framework in terms of key performance metrics in ICSNs viz., average in-network delay, and publisher load, given different mobility pause time durations and data consumers' densities.

In Chapter 5, we elaborated more on the learning techniques that could be used with AHP-based reasoning to improve the cognitive decision making capabilities at the local cognitive nodes of the ICSN, operating in an IoT application environment. The challenge was to adapt the learning to respond to varying user requirements/requests, to consider the energy consumption in the network while catering to the QoI requirements, and to maintain a high average rate of a successful data delivery to the sink while doing so. To address these challenges, we proposed two learning techniques, namely, learning data delivery A* (LDDA*) and cumulative-heuristic accelerated learning (CHAL) which employ heuristics towards improving the data delivery rate. LDDA* was able to deliver data with good QoI at the sink, while CHAL was the more energy-considerate technique, and both of these learning strategies had comparable performance with respect to the success rate of data delivered to the sink. Toward the end of this chapter, extensive simulations had shown improvement of about 40 percent in the average rate of successful data delivery to the sink with the use of heuristic learning, when compared with a network that didn't implement any learning.

In Chapter 6, we proposed a framework for data delivery in large-scale networks for disaster management, where numerous wireless sensors are distributed over the city traffic infrastructures, shopping mall parking areas, airport facilities, and so on. In general, our framework catered for energy-efficient applications in the IoT where data is propagated via relays from diverse sensor-nodes toward a gateway connected to a large-scale network such as the Internet. We considered the entire network energy while choosing the next hop for the routed packets in the targeted wireless sensor network. Our delivery approach considers resource limitations in terms of the hop count, and remaining energy levels. Moreover, the achieved results confirmed the effectiveness of the proposed approach in comparison to other baseline energy-aware routing protocols in the literature.

In Chapter 7, we proposed an adaptive routing approach (ARA) that selectively launches routes of communication between the heterogeneous IoT nodes. Since nodes in the IoT belong to different owners, we also introduced a pricing mechanism to cater to the exchange of monetary costs by intermediate nodes to utilize their relaying resources. In this chapter, the ARA routing approach was verified via a case study, in addition to theoretical analysis while demonstrating the utility and practicality of ARA in the heterogeneous IoT as it scales.

In Chapter 8, we proposed a cache replacement approach based on the value of sensed information (VoI) in one of the future IoT trends, namely the fog applications. Our approach depended on three functional parameters in information-centric sensor networks (ICSNs). These three parameters were the age of data based on a periodical request, the popularity of the on-demand requests, and the duration for which the sensor node is required to operate in active mode to capture sensed readings. These parameters are considered together to assign a value to cached data in fog to retain the most valuable information in the cache closer to the end user for prolonged time periods. This strategy provided significant availability for most valuable and difficult to sense data in the ICSNs. In addition, simulation results were performed to compare it against other dominant cache replacement policies under varying circumstances such as data popularity, cache size, network load, and connectivity degree. Potential and promising results were achieved accordingly.

9.2 Future Directions

With the research done in this book, ICSNs will be able to provide better infrastructure support for the Smarter Planet initiatives across the globe. Not only this, several future research directions and open issues can also be derived from the work done in this book so far.

In the following, we outline some of these future directions and open research issues.

1. Exploiting the caching capabilities more and more.

    One of the aspects of the information-centric capabilities in the ICSN framework, is its ability to cache information in cognitive nodes and use it collaboratively for information sharing across the network. While we acknowledged this advantage, such caching would rise an issue with the mobility-enabled node presence, which we did not delve deeply into yet. Thus, exploring the role of caching in information access and data delivery, and the study of cache replacement techniques that suit the cognitive nodes in mobile ICSNs is still a direction to explore out of this work.

2. Mobility-enabled cognitive nodes.

    In general, the effect of node mobility on the QoI, and its impact on the ICSN adaptability and longevity shall be investigated more for safer ITS systems in the near future. While the discussed cognitive ICSN architecture proposed here can help in overcoming the limitations of the cross-layer design, we strongly recommend further investigations in the physical layer components and configurations.

3. More developed and commercialized cognitive gateways.

    In terms of enhancing the cognitive gateway while supporting diverse application platforms, this work can be reconsidered for further enhancement in domain-specific ontologies for better knowledge representation. The creation of such ontologies can contribute toward the development of an enterprise architecture that can be applied to different application domains using the same underlying cognitive sensor network.

4. ICSNs integration with the next generation wired/wireless networks.

    More functions could be incorporated at the cognitive gateway to integrate it with the next generation networks (NGNs) toward more cognitive radio enable nodes, working in cognitive network setups. The expansion of the gateway functions would then be able to take requests directly from different network users such as cell phones, access points, and base stations, thus making the ICSN platform more accessible to end users in an IoT paradigm.

5. Cognition in the future Internet.

   The idea of cognition can be investigated more while being applied to intermediate routers/switches of the current Internet infrastructure to realize the cognitive network concept in general. Data need not be requested from specific hosts, nor has to travel end-to-end in the network. Instead, it can be cached at the network's edge (fog). Cognitive routers could be used to understand the user request patterns and manage the cache content intelligently.

6. Energy optimization.

   Researchers shall investigate and look for an optimal learning policy that will identify the minimum CNs required while minimizing the number of connected sensors to it. The minimum cover set approach with optimal learning can be used to find the minimum number of sensor nodes, to minimize energy consumptions without violating the networked information quality/accuracy. CNs can communicate with each other and share their knowledge toward a common network goal. This information sharing can speed up the network learning, and thus can make better decisions about the ICSN resource management much earlier. This can lead to dramatic improvements in energy savings.

7. Flexibility of the cognitive network framework.

   The advantage of using a learning mechanism such as the AHP-based one stems mainly from applying the machine learning theory to measurable QoI attribute(s) in the wireless/wired IoT network, and thus making the cognitive ICSN framework adaptable to numerous IoT applications. The priorities assigned to the QoI attributes could also be varied to influence the classification of traffic flows, and the choice of the data delivery paths. This shall be more emphasized in the future work of this book.

8. Security and privacy open issues.

   Security and privacy are key issues to be addressed while spreading the adoption of cognitive networks in IoT applications. Unluckily, security solutions have not been incorporated in a planned way in the early IoT version. The enabling technologies of RFID and sensor networks were simply integrated across with the existing Internet infrastructure, without considering key design factors such as the trust and privacy issues while gathering and delivering data. These security considerations span not only along the data management, application, and service levels, but also the communication and networking levels, and are significant aspects to be investigated in any future work.

Index